Late Cenozoic Yushe Basin, Shanxi Province, China: Geology and Fossil Mammals

Vertebrate Paleobiology and Paleoanthropology Series

Edited by

Eric Delson
Vertebrate Paleontology, American Museum of Natural History,
New York, NY 10024, USA
delson@amnh.org

Eric J. Sargis
Anthropology, Yale University
New Haven, CT 06520, USA
eric.sargis@yale.edu

Focal topics for volumes in the series will include systematic paleontology of all vertebrates (from agnathans to humans), phylogeny reconstruction, functional morphology, Paleolithic archaeology, taphonomy, geochronology, historical biogeography, and biostratigraphy. Other fields (e.g., paleoclimatology, paleoecology, ancient DNA, total organismal community structure) may be considered if the volume theme emphasizes paleobiology (or archaeology). Fields such as modeling of physical processes, genetic methodology, nonvertebrates or neontology are out of our scope.

Volumes in the series may either be monographic treatments (including unpublished but fully revised dissertations) or edited collections, especially those focusing on problem-oriented issues, with multidisciplinary coverage where possible.

For other titles published in this series, go to
www.springer.com/series/6978

Late Cenozoic Yushe Basin, Shanxi Province, China: Geology and Fossil Mammals

Volume I: History, Geology, and Magnetostratigraphy

Edited by

Richard H. Tedford (1929–2011, deceased)

Formerly Division of Paleontology, The American Museum of Natural History, Central Park West at 79th Street, New York, NY 10024-5192, USA

Zhan-Xiang Qiu

Laboratory of Paleomammalogy, Institute of Vertebrate Paleontology and Paleoanthropology, Chinese Academy of Sciences, Xizhimenwai Ave., 142, Beijing 100044, People's Republic of China

Lawrence J. Flynn

Department of Human Evolutionary Biology, and Peabody Museum of Archaeology and Ethnology, Harvard University, Cambridge, MA 02138, USA

 Springer

Editors
Richard H. Tedford
Formerly Division of Paleontology
The American Museum of Natural History
New York
NY, USA

Zhan-Xiang Qiu
Laboratory of Paleomammalogy
Institute of Vertebrate Paleontology
  and Paleoanthropology
Chinese Academy of Sciences
Beijing
People's Republic of China

Lawrence J. Flynn
Department of Human Evolutionary Biology,
  and Peabody Museum of Archaeology
  and Ethnology
Harvard University
Cambridge
MA, USA

ISSN 1877-9077   ISSN 1877-9085   (electronic)
ISBN 978-90-481-8713-3   ISBN 978-90-481-8714-0   (eBook)
DOI 10.1007/978-90-481-8714-0
Springer Dordrecht Heidelberg New York London

Library of Congress Control Number: 2013930357

Cover Illustration: Exposure of Yushe Group in Zhangcungou, one of Licent's fossil localities. The steep wall situated right, and below the level of the trees in the centre of the picture, is the place where Licent's only profile was measured (Licent and Trassaert, 1925, Fig. 1). Photo taken August 9, 1997, by Z.-X. Qiu

Printed on acid-free paper

Springer is part of Springer Science+Business Media (www.springer.com)

Foreword

This is the first volume of *Late Cenozoic Yushe Basin, Shanxi Province, China: Geology and Fossil Mammals*, a series to be published by Springer on the terrestrial vertebrate fauna of the last six and a half million years from an area of North China that figured regionally in the development of modern vertebrate paleontology. The geological and paleontological record is important for knowledge of the Northeast Asian terrestrial fauna of the Late Neogene, and for the history of science in China. Yushe strata record in one place a succession of faunas that for many years constituted the best body of evidence on the paleobiology of Asia through all of Pliocene time. It still stands as perhaps the best place to document the succession of earlier Pliocene mammalian faunas in China. The present volume is an introduction to Yushe. Its focus is the geology of Yushe Basin, the history of exploration there, and the magnetostratigraphy that, in concert with the fossil record, yields a stratigraphy with a well-resolved time frame.

Volume I of the Late Cenozoic Yushe Basin Series, "*History, Geology and Magnetostratigraphy*," is to be followed by sequential studies on the systematics of mammalian groups, with emphasis on their provenance. We plan five additional volumes to follow on phylogenetically and geologically documented investigations of distinct mammalian groups. These data may then be applied to produce a refined, composite biostratigraphy for Yushe Basin, which can be used for correlation elsewhere and for interpretation of faunal events, including turnover, appearances, and dispersals. The interpretive and synthetic biochronology and paleobiogeography will be integrated into the final volume.

We dedicate this volume especially, and the whole series on the paleontology of Yushe Basin, to Richard H. Tedford (1929–2011). Volume I of *Late Cenozoic Yushe Basin, Shanxi Province, China: Geology and Fossil Mammals* was born of the collaboration that began in 1981 between Dick Tedford and Zhan-Xiang Qiu. During the 1990s, Dick Tedford and Zhan-Xiang Qiu compiled their geological and stratigraphic observations, and in collaboration with Neil Opdyke, University of Florida, built the temporal framework for interpretation of the fossil record of Yushe. Progress on the entire, diverse mammalian fauna followed, but serious health issues for both Dick and Zhan-Xiang were a setback to the project in 2003. In recent years, work continued at a diminished pace. Dick was able to complete his interpretation of the geology of Yushe Basin in 2010, even after his health once again began to deteriorate. His passing in July of 2011 is difficult for all of us and a loss for the body of knowledge of Neogene terrestrial faunas from around the world. This volume in particular, as well as the whole series of Springer volumes, was the vision of Dick Tedford, and he contributed in many ways to the research on all of the Yushe vertebrates treated in subsequent works. We miss his genuine passion for understanding of the vertebrate record, his interest in all groups, and his encouragement to develop a refined biostratigraphy that would serve as a standard for biochronology in Northeastern Asia.

April 2012   Zhan-Xiang Qiu
                                                                              Lawrence J. Flynn

Preface

Why don't you try to be another "Teilhard de Chardin"? This was the question Qiu ardently addressed to Tedford, staring directly into his face, glowing under the golden light of the half-set sun. Both of us were sitting on the slope of a big block of Permo-Triassic violet sandstone after a day of field work in the Yushe Basin in early summer, 1982. Tedford did not answer the question, but immersed himself in deep meditation. All the things that happened thereafter may take their origin from this short unanswered question. Why? And why not?

For Qiu, he felt deeply the need to find a paleontologist rivaling Teilhard de Chardin in scope and depth of knowledge to resume the work in Yushe after almost 40 years of stagnation. The importance of Yushe as a key area for Late Neogene stratigraphy and vertebrate paleontology had been well understood by all the Chinese students studying Neogene problems. In the 1930s, when rich mammalian fossils were first found in the Yushe area, primary importance was attached to the fluviatile-lacustrine character of fossil-bearing deposits radically distinct from the wide-spread "*Hipparion* Red Clay" of North China. It was soon recognized that the main body of the Yushe fossils belonged to an unknown fauna intermediate between the Baode *Hipparion* fauna and the Nihewan *Equus* fauna. Teilhard de Chardin undertook research work on the Yushe mammalian faunas, but he failed to finish this research before he left China in 1945. This left the full nature of one of the most important late Cenozoic faunas from being clearly revealed. This probably was one of the reasons why Prof. C. C. Young, the late founder of the Institute of Vertebrate Paleontology and Paleoanthropology (IVPP), who visited Yushe himself during the 1930s, sent a team to resume the Yushe studies in 1955–1956 soon after he was reappointed as the Director of the Cenozoic Laboratory. When Qiu was transferred from the Paleogene Division of the Laboratory of Paleomammalogy, IVPP, to the Neogene Division, Yushe was naturally chosen as his first target. For Tedford, his fascination with the Chinese Neogene and its mammalian faunas came from his acquaintance with the marvelous specimens collected from China, and then housed in the Frick Collection at the American Museum of Natural History (AMNH), which became fully accessible to him in 1968.

These common interests united them. In 1981, when they first met in the "Horse Heaven" workshop at the AMNH, strong desires from both sides led to a plan to initiate an intimate cooperation on the Chinese Neogene stratigraphy and mammalian faunas. The outcome led to the above scene that occurred in the following year. However, other commitments and the need to raise grant money delayed initiation of this joint project until 1987. In 1982–1984, Qiu went to Germany on a Humboldt scholarship to enhance his knowledge of the European Neogene, while Tedford continued preparation for the Yushe Project. Tedford succeeded in persuading Neil Opdyke, the leading authority on terrestrial paleomagnetics; Larry Flynn, a small mammal expert with focus on the Asian Neogene; and Will Downs, whose rich experience in field work and his ability in reading and speaking Chinese turned out to be extremely valuable, to join the project from the American side. Meanwhile, Qiu recruited Wu Wenyu, a leading figure in micro-mammalogy in China, Yu-Qing Li, De-Fa Yan, Guan-Fang Chen, Jie Ye, and Wei Dong, all very active large mammal specialists, and a number of postgraduate students, Xiao-Feng Chen, Yi-Zheng Li, and Gen-Zhu Zhu, to join the project from the Chinese side.

At the very beginning of the joint project, the aims were set forth clearly. (1) To pursue research on the large mammalian fossils of Yushe left unstudied (Canidae, Ursidae, Rhinocerotidae, Suidae, part of Hyaenidae, and part of Equidae); (2) To revise the groups already studied in light of the current knowledge of their systematics; (3) To obtain as many small mammals as possible from stratigraphically defined layers to overcome the strong bias toward large mammals in the early collections; (4) To assign as many mammalian fossils as possible to a solid stratigraphic foundation; (5) To apply modern techniques and methods in field work, especially wet sieving for small mammalian remains; and (6) To combine magnetostratigraphy with biostratigraphy in order to date the faunal succession.

Large-scale fieldwork began in 1987 and concluded in 1991. Subsequently, a number of short trips aimed at solving specific geological problems in Yushe Basin were undertaken by a few team members. Work continued on faunal remains but suffered a few setbacks. De-Fa Yan's key work on ungulates was cut short prematurely, but continued by collaborators. Also unfortunately, Will Downs died toward the end of 2002, but his contribution of translations from the Chinese literature continues to provide insight not only to our project, but also for other non-Chinese writers becoming interested in diverse Chinese earth sciences. Will was a key player in developing the microfaunal record of Yushe, which includes a densely sampled view of the Late Neogene small mammal communities of Shanxi Province. Will's enthusiasm and fascination with China and his contributions to our work provide a lasting gift to us and colleagues of similar interest.

This volume presents the results obtained from both field and laboratory work. As we were preparing the Yushe series of volumes, great progress in the field of Neogene mammalogy and stratigraphy was being made both in other places in China and elsewhere in Eurasia. The Yushe Basin remains the most important and most informative Late Neogene basin in China. It provides the most complete section from latest Miocene through the entire Pliocene and into the Early Pleistocene, with only modest gaps, and it is paleomagnetically calibrated and paleontologically dated based on rich small and large mammalian faunas.

Zhan-Xiang Qiu

Acknowledgments

We take this opportunity to express our deep gratitude to all the institutions and persons who warmly supported the Yushe Project in various ways. For financial support, we thank the National Science Foundation (NSF) of the USA and the National Natural Science Foundation (NSFC, China), and the Chinese Academy of Sciences. Two NSF grants supported the field work: EAR 8709221 and BSR 9020065; later synthesis was supported under Grant No. 0958178. For manifold material and spiritual support, we thank the authorities and our colleagues of the American Museum of Natural History, the Institute of Vertebrate Paleontology and Paleoanthropology, the Tianjin Natural History Museum, and the Naturhistorisches Museum, Basel. Last, but not least, we thank the local authorities and the people of Yushe County, without whose help the present work could not have been so successfully completed.

We dedicate this volume to our friend and colleague Dick Tedford, and at the same time we recognize all those who went before us during the past century, on whose work our conclusions and hopeful advancements depend so heavily. Some of those were in the field with us and inspired us by their dedication, including De-Fa Yan and Will Downs. Volume I was assembled thanks to the contributions of many colleagues and associates, most recently by the efforts of members of IVPP in Beijing and staff members of AMNH in New York, especially Alejandra Lora, Judy Galkin, and Frank Ippolito. Frank's skill in preparing artwork, envisioned by Dick Tedford, is technically flawless and artistically admirable. Susan Bell, Ruth O'Leary, and Loraine Meeker aided this effort in diverse ways, making access to key documents possible.

We thank Judith Terpos and of the entire Springer staff for their help in realizing the first of the Yushe Basin volumes. The meticulous efforts of series editors Eric Sargis and Eric Delson, are deeply appreciated. They encouraged the maturation of the final product and its artwork. The more recent constructive help of external manuscript reviewers is greatly appreciated. Their careful reading led to various improvements in clarity throughout the volume.

Vivien Tedford and Andrew Woodford assisted Dick in completing his work on this volume, and continued to help us assemble parts of the present manuscript through the summer of 2011. Richard H. Tedford made this volume possible by his charisma and leadership, and by his example that shows what constitutes good science.

Contents

1 Yushe Basin, Shanxi Province.. 1
Zhan-Xiang Qiu and Lawrence J. Flynn

2 History of Scientific Exploration of Yushe Basin.................... 7
Zhan-Xiang Qiu and Richard H. Tedford

3 Cenozoic Geology of the Yushe Basin............................ 35
Richard H. Tedford, Zhan-Xiang Qiu and Jie Ye

4 The Paleomagnetism and Magnetic Stratigraphy of the Late Cenozoic
Sediments of the Yushe Basin, Shanxi Province, China............... 69
Neil D. Opdyke, Kainian Huang and R. H. Tedford

5 Biostratigraphy and the Yushe Basin............................ 79
Lawrence J. Flynn and Zhan-Xiang Qiu

6 Erratum To: The Paleomagnetism and Magnetic Stratigraphy of the Late
Cenozoic Sediments of the Yushe Basin, Shanxi Province, China......... E1
Neil D. Opdyke, Kainian Huang and R. H. Tedford

Appendix I: Andersson's Localities in Southern Shanxi Province 83

Appendix II: Licent's Trips to Southeastern Shanxi.................... 85

Appendix III: Excerpts from Licent's Diary
           (Vide supra, Fig. 2.7, Route B)........................ 87

Appendix IV: Licent's Localities in Southeastern Shanxi
           (Vide supra, Fig. 2.7)................................. 93

Appendix V: Excerpts from the Correspondence
           of E. Nyström and C. Frick............................ 97

Appendix VI: Quan-Bao Gan's Collection from Yushe................. 99

Appendix VII: Some of Zhan-Xiang Qiu's Collection from the Yushe
           Area, 1979–1981.................................... 101

Appendix VIII: Columnar Sections Measured by the Sino-American Yushe Project   103

Appendix IX: China–America Yushe Field Party Members and Contributors to the Series Late Cenozoic Yushe Basin, Shanxi Province, China: Geology and Fossil Mammals   105

Index   107

Contributors

Lawrence J. Flynn
Department of Human Evolutionary Biology, and Peabody Museum of Archaeology and Ethnology, Harvard University, Cambridge, MA 02138, USA
ljflynn@fas.harvard.edu

Kainian Huang
Department of Geological Sciences, University of Florida, Gainesville, FL 32611, USA
knhuang@ufl.edu

Neil D. Opdyke
Department of Geological Sciences, University of Florida, Gainesville, FL 32611, USA
drno@ufl.edu

Zhan-Xiang Qiu
Laboratory of Paleomammalogy, Institute of Vertebrate Paleontology and Paleoanthropology, Chinese Academy of Sciences, Xizhimenwai Ave., 142, Beijing 100044, People's Republic of China
qiuzhanxiang@ivpp.ac.cn

Richard H. Tedford
Formerly Division of Paleontology, The American Museum of Natural History, Central Park West at 79th Street, New York, NY 10024, USA

Jie Ye
Laboratory of Paleomammalogy, Institute of Vertebrate Paleontology and Paleoanthropology, Chinese Academy of Sciences, Xizhimenwai Ave., 142, Beijing 100044, People's Republic of China
yejie@ivpp.ac.cn

Chapter 1
Yushe Basin, Shanxi Province

Zhan-Xiang Qiu and Lawrence J. Flynn

Abstract Yushe Basin is an intermontaine basin located in northern China. It lies at the eastern edge of the Loess Plateau, just west of the Taihang Mountains. Its fluvial, lake and, finally, loess deposits accumulated during the last 6.5 Myr and contain many fossiliferous horizons. As a site for early scientific explorations, Yushe figures into the history of the development of vertebrate paleontology in China over the last century. We were able to relocate many of the early twentieth century fossil localities of Yushe, and add significant new paleontological discoveries. Fossils document Late Miocene assemblages, terrestrial faunas characteristic of North China during most of the Pliocene, and an Early Pleistocene community comparable to that of Nihewan Basin. The succession of Yushe faunas spanning the Pliocene Epoch is unsurpassed elsewhere in China in richness and depth of time covered. The Pliocene assemblages characterizing the Yushean chronofauna cluster in two successive units, and provide the basis for characterizing the Gaozhuangian and Mazegouan land mammal stage/ages. This volume documents the geological context of the rocks and faunas of Yushe and provides the justification for age determination of the fossiliferous deposits.

Keywords Yushe Basin • Late Neogene • Biochronology • Pliocene • North China • History of science

1.1 Physical and Historical Setting

The Yushe Basin has produced key terrestrial vertebrate faunas over the last century and has thus played an important role in the growing twentieth century knowledge of the nature of successive mammalian faunas of Pliocene age in northern Asia. Early in that century and systematically so in the 1920s, scientists began exploring China for its vertebrate fauna. The Cenozoic Research Lab was established in 1929 as a subordinate of the Geological Survey of China, and was a center of international collaboration, especially for Chinese and European investigators. Important field areas in Shanxi Province included Yushe and Wuxiang to the south, as well as Baode (Kurtén 1952), some 300 km to the northwest (Fig. 1.1). Pliocene and Late Miocene deposits are widespread in Shanxi Province, and early exploration extended into the Shouyang area, east of the capital of the province, Taiyuan. It was clear early on, and abundantly clearer in recent years (Qiu 1987), that Yushe and other areas in Shanxi Province offered a laboratory in which to unravel mammalian biostratigraphy for North China on a fine scale, and with a succession of vertebrate faunas.

Yushe is a town in southern Shanxi Province. In the valley of the Zhuozhanghe (the suffix "he" means "river"), Yushe farmland is developed on Late Neogene terrestrial sediments. These are fluvial and lacustrine sediments that accumulated in a complex of five related subbasins. Collectively (Fig. 1.2), the Yuncu, Nihe, Ouniwa, Tancun, and Zhangcun subbasins comprise Yushe Basin. The subbasins are generally separated by Triassic bedrock, but apparently are connected in some cases, calling for further structural analysis of the basement. Nonetheless, all yield vertebrates of Pliocene age, and the adjacent Yuncu and Tancun subbasins contain older Late Miocene basin fill with

Z.-X. Qiu (✉)
Laboratory of Paleomammalogy, Institute of Vertebrate Paleontology and Paleoanthropology, Chinese Academy of Sciences, Xizhimenwai Ave., 142, Beijing 100044, People's Republic of China
e-mail: qiuzhanxiang@ivpp.ac.cn

L. J. Flynn
Department of Human Evolutionary Biology and Peabody Museum of Archaeology and Ethnology, Harvard University, Cambridge, MA 02138, USA
e-mail: ljflynn@fas.harvard.edu

R. H. Tedford, Z.-X. Qiu, L. J. Flynn (eds.), *Late Cenozoic Yushe Basin, Shanxi Province, China: Geology and Fossil Mammals. Volume I: History, Geology, and Magnetostratigraphy*, Vertebrate Paleobiology and Paleoanthropology, DOI: 10.1007/978-90-481-8714-0_1, © Springer Science+Business Media Dordrecht 2013

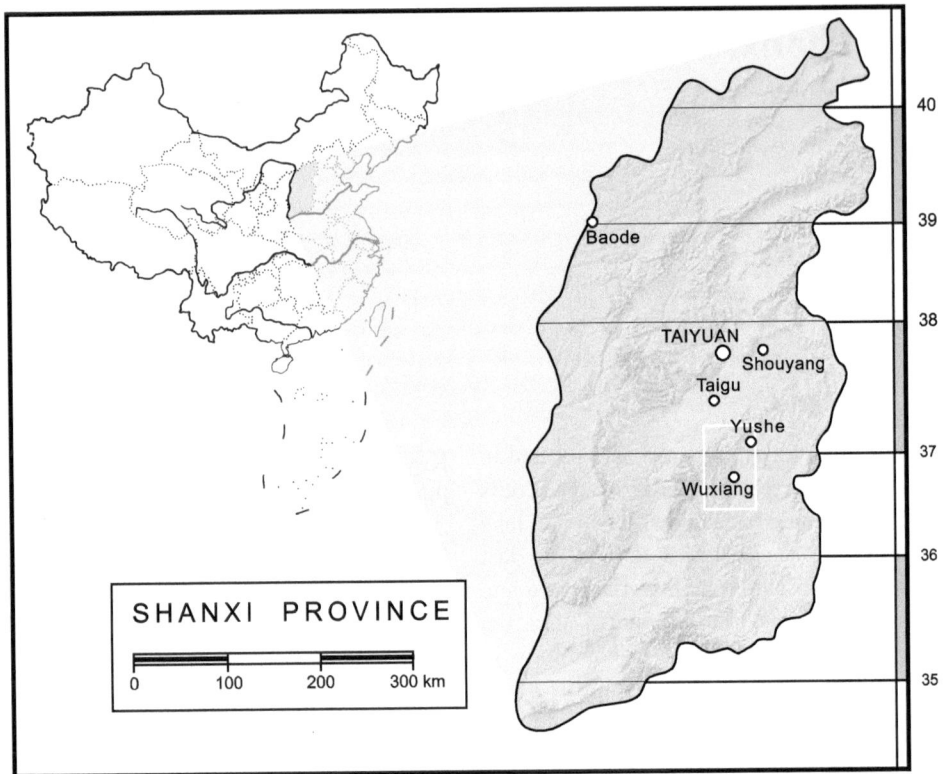

Fig. 1.1 Location of Shanxi Province within North China, indicating the Yushe-Wuxiang area shown in Fig. 1.2, the important fossil areas of Baode and Shouyang, and the provincial capital Taiyuan, as well as Taigu City, which was an important departure point and shipping station in the early days of collecting. This figure and the next were conceived by Dick Tedford and skillfully interpreted by Frank Ippolito (AMNH)

fossils as well. South and west of Yushe Basin is another basin complex, the Wuxiang Basin, which yields fossils of similar age.

1.2 The Sino-American Yushe Project

Zhan-Xiang Qiu and Richard H. Tedford mounted a collaborative Sino-American investigation of the geology and fossil mammals of Yushe Basin. Beginning in 1987 they conducted a series of joint field campaigns to measure sections and tie fossil occurrences unambiguously into the composite biostratigraphy. Their team of collaborators was able to relocate classical localities quarried in the early 1900s and add numerous new fossil occurrences, including small mammals, which had been underrepresented previously. Figure 1.3 is an example of the yearly (at least in early years) group photograph, assembling visiting geologists both from IVPP and abroad with local paleontologists and interested public.

The focus of our study was Yuncu subbasin stratigraphy and biostratigraphy. Paleontologically, we analyzed all fossils from Yushe, but it is the Yuncu stratigraphy, supplemented by information from Tancun subbasin, that

formed the basis for our biostratigraphic observations and biochronologic conclusions. Figure 1.2 indicates the areas of more detailed geological mapping presented in Chap. 3. Neil Opdyke sampled the Yuncu section for paleomagnetic analysis. His correlation (Chap. 4) of the magnetostratigraphy to the Geomagnetic Polarity Time Scale (GPTS) refines the dating indicated by biochronology of the sediments and faunas of Yushe Basin. Yushe strata and fossil horizons span an interval from about 6.5 to 2 Ma, the Late Miocene, entire Pliocene, and earliest Pleistocene epochs. The younger loess also contains fossils.

Our Yushe project localities can be plotted on the composite biostratigraphy, as depicted in the last figure of the magnetostratigraphy chapter. Their age relationships are well resolved. On the other hand, localities of historic importance for fossil occurrences developed by Licent, the Frick collectors, and later IVPP also can be placed in the sequence, in some cases with precision, as a result of our work. Hence, the results of the Sino-American collaboration yield a biostratigraphic data set crucial to understanding events of the Late Neogene in Asia, in the interval of ∼6.5 to ∼2 Ma. We use the chronological unit "Neogene" in this volume as a convenient way of grouping later Cenozoic Era time. This unit of period magnitude includes

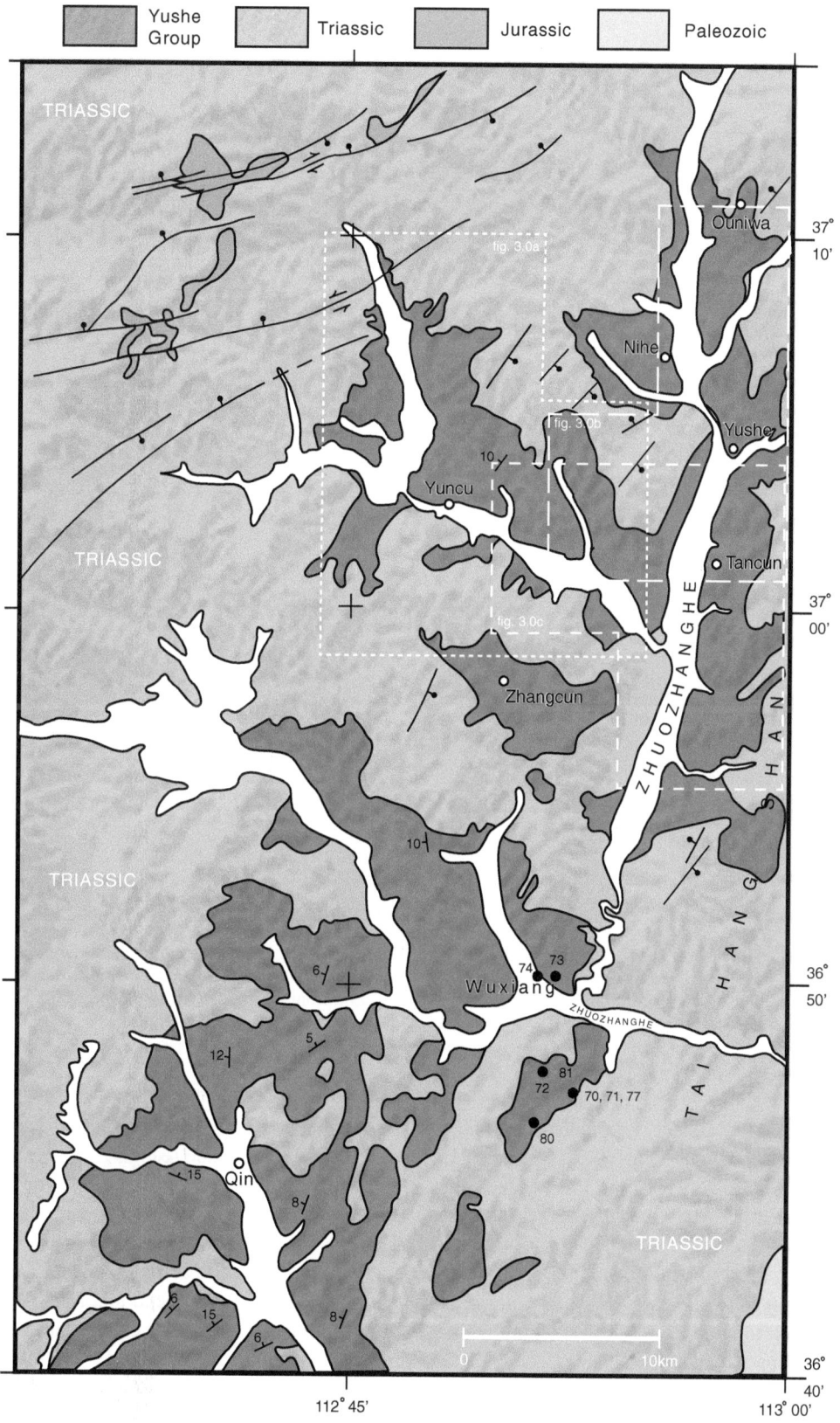

Fig. 1.2 Geological map of the Yushe-Wuxiang area, showing the distribution of Yushe Group sediments in the Yushe Basin and Wuxiang Basin. Yushe Basin comprises five subbasins, Yuncu, Tancun, Nihe, Ouniwa, and Zhangcun. The *white dashed rectangles* outline the areas of more detailed geological maps discussed in Chap. 3 (Fig. 3.1a–c). *Small numerals* denote local dip; *large filled dots* and *numerals* are Andersson field localities

Fig. 1.3 Group photo at the entry to the then-new Yushe County Museum. On the *far left* in the *front row* is Tai-Ming Wang of the Yushe County Museum, *fourth* from *left* is De-Fa Yan of IVPP, followed by Will Downs, Dick Tedford, and Zhan-Xiang Qiu. Wen-Liang Jia of the Yushe County Museum is *second* from *right*, and Wen-Yu Wu of IVPP is at the *right*. In the *back row* are three then-new post-graduates of IVPP, Yi-Zheng Li, Xiao-Feng Chen, and Gen-Zhu Zhu from *left* to *right*, with Ye Jie of IVPP behind Tedford and Qiu (photo by L. Flynn, September 5, 1987)

the Miocene, Pliocene, Pleistocene, and Holocene epochs. We use "Late Neogene" without defining it as a shorthand reference to Late Miocene, Pliocene, and Pleistocene time, approximately the last 11 Myr. For us, Late Neogene applies to all of the Yushe Group sediments.

Herein, we tend to treat biostratigraphic results in terms of the time scale resolved to 10^5 years. For both intra-basin studies and correlation to other dated locales, we utilize the calculated ages of fossils and assemblages, rather than inferred marine or biochron time unit correlates. In many contexts, therefore, the corresponding marine stage or MN unit, or even the choice of placement of the Pliocene–Pleistocene boundary is irrelevant, because we can apply a numerical age. As to the beginning of the Pleistocene, we follow the consensus that apparently prevails (see, e.g., Mascarelli 2011), which recognizes the boundary at 2.6 Ma. This means that upper Yushe levels (the Yuncu subbasin Haiyan Formation) are considered Early Pleistocene, although the magnetostratigraphy is pre-Olduvai and previously was referred to as Late Pliocene (e.g., Tedford et al. 1991). It is important to recall that North China experienced faunal turnover at around 2.6 Ma (and Yushe biostratigraphy reflects this; Flynn et al. 1991), so in the context of intensified faunal change, consideration of upper Yushe levels as Early Pleistocene in age is useful conceptually.

The present book is the initial volume in the Springer series *Late Cenozoic Yushe Basin, Shanxi Province, China: Geology and Fossil Mammals*, Volumes I–VI. Volume I on the geology and history of exploration of Yushe Basin gives the magnetostratigraphic framework and introduces systematic studies on the vertebrate fossils of the basin. Subsequent volumes treat the fossil fauna in systematic units. Volume II concentrates on small mammals, primarily rodents and lagomorphs, but also insectivores and bats. That work draws on classic studies (e.g., Teilhard de Chardin 1942) of larger fossils picked up while prospecting exposures in the early part of the last century, but it is based mainly on new collections made during our field campaigns of the late 1980s and early 1990s. We used the technique of wet screening individual localities, and these were excavated by us with reference to the magnetostratigraphic sections. Therefore, for most small mammals, stratigraphic provenance can be given precisely.

Volume III will present the diverse Carnivora and rare Primates of Yushe Basin. This collaborative effort includes key work by Dick Tedford on many chapters. It draws on collections from classic localities, such as those discovered by E. Licent, and currently housed in several institutions, and on fossils found in recent years. Volumes IV, V, and VI concern the major groups Perissodactyla, Artiodactyla, and Proboscidea, in part studied in previous years. New work brings these groups up-to-date. The series closes with coverage of remaining taxa, a synthesis of the biostratigraphy of Yushe, and a treatment of the biogeography of the various groups, which illustrates east–west communication with western Asia and Europe, but also with North America. Faunal connections to the south were much more restricted. All of these issues are treated in the later volumes of the series.

The Sino-American Yushe Project (SAYP) is multidisciplinary and involves specialists from diverse fields (see appendix). These collaborators coauthor the chapters of the volumes that follow the present work on the geology, history of exploration, and magnetostratigraphy of Yushe. We attempt to standardize citation of chapter authors by placing the family name second for all. Traditionally, of course, Chinese family names come first. At the same time as inverting the Chinese name order, the given names are usually compound: hyphenated, with the second syllable capitalized. There are exceptions: some given names are monosyllabic and some individuals do not hyphenate their

names. Invariably, however, citations of references give family names.

Yushe Basin played an important role in the history of paleontology of China, indeed of Asia. The fauna from diverse localities—and ages—in the basin was recognized to document Pliocene mammalian life in northeastern Asia, and it remains one of the best records of Late Neogene paleofaunas in the region. In contrast to single sites with rich assemblages, Yushe localities together document the faunal history of the region in a continuous record spanning approximately 6.5–2 Ma.

The Springer series of volumes on the *Late Cenozoic Yushe Basin, Shanxi Province, China: Geology and Fossil Mammals* offers in one source the results of a modern collaborative effort between institutions based in several countries. Our collaboration was extremely fruitful and demonstrated that control of the context of fossil provenance, coupled with dating, yields a biostratigraphic data set that can be applied to the study of past faunas as paleobiological entities, and to the study of how faunal assemblages change through time and at what rate. As such, this Yushe Basin analysis applies to research questions on the origin of the modern Asian temperate fauna, an appreciation of past diversity, and caution for modern peoples on the fragility of ecosystems and the perils of extinction.

References

Flynn, L. J., Tedford, R. H., & Qiu, Z.-X. (1991). Enrichment and stability in the Pliocene mammalian fauna of North China. *Paleobiology, 17*(3), 246–265.

Kurtén, B. (1952). The Chinese *Hipparion* fauna. *Commentationes Biologicae, 13*, 1–82.

Mascarelli, A. L. (2011). Quaternary geologists win timescale vote. *Nature, 459*, 624.

Qiu, Z.-X. (1987). Die Hyaeniden aus dem Ruscinium und Villafranchium Chinas. *Münchener Geowissenschaftliche Abhandlungen, A9*, 1–110.

Tedford, R. H., Flynn, L. J., Qiu, Z.-X., Opdyke, N. D., & Downs, W. R. (1991). Yushe Basin, China: Paleomagnetically calibrated mammalian biostratigraphic standard for the Late Neogene of eastern Asia. *Journal of Vertebrate Paleontology, 11*(4), 519–526.

Teilhard de Chardin, P. (1942). New rodents of the Pliocene and Lower Pleistocene of North China. *Publications de l'Institut de Géobiologie, Pékin, 9*, 1–101.

Chapter 2
History of Scientific Exploration of Yushe Basin

Zhan-Xiang Qiu and Richard H. Tedford

Abstract The Yushe Basin is a complex of subbasins that accumulated fluvio-lacustrine deposits. The Yuncu subbasin contains the longest, most complete stratigraphic record and yields abundant fossils from multiple horizons. Vertebrate paleontology began in China early in the twentieth century, and since the 1920s Yushe Basin figured prominently in recognition of pre-Pleistocene faunas of North China. Emile Licent's collecting campaign of 1934–1935 established the richness of Yushe Basin's fossils and stimulated later exploration by collectors for Childs Frick of the American Museum of Natural History. Many new taxa were based on Yushe Basin fossils, but field work languished through much of the last century. Renewed interest in Yushe led to the formal field project conducted jointly by a Sino-American team in the 1980–1990s. This field work documented the localities of early finds, analyzed the composite section and dated it by paleomagnetic reversal stratigraphy, and added important new fossil data.

Keywords Yushe Basin • North China • History of paleontology • Miocene • Pliocene • Pleistocene • Vertebrate fossils • Collaborative research

Z.-X. Qiu (✉)
Laboratory of Paleomammalogy, Institute of Vertebrate Paleontology and Paleoanthropology, Chinese Academy of Sciences, Xizhimenwai Ave., 142, Beijing 100044, People's Republic of China
e-mail: qiuzhanxiang@ivpp.ac.cn

R. H. Tedford
Formerly Division of Paleontology, American Museum of Natural History, Central Park West at 79 St, New York, NY 10024, USA

2.1 Geographical Setting

The Yushe Basin is situated in the northern part of the low mountainous area of southeastern Shanxi Province. This mountainous area is bordered by two sub-parallel mountain ranges (shan) stretching in a NNE to SSW direction: the Taihang Shan to the east, and the Taiyue Shan to the west. The Taihang Shan are about 550 km in length, merging with the Yan Shan mountain ranges on the north and ending before reaching the Huang He (Yellow River) valley to the south, turning westwards and connecting with the Zhongtiao Shan. The highest peak of the Taihang Shan within Shanxi Province is 2097 m (Beitianchi Peak) above sea level. It also serves as the western border of the North China Plain. The Taiyue Shan are shorter but wider, and rather irregular in form, with their highest peak (Niujiao'an) being 2568 m above sea level. West of the Taiyue Shan is the Fen He (Fen River) graben, which is generally 300 m lower than the bottoms of the basins in the mountainous area of southeastern Shanxi (Fig. 2.1). To the north, the mountainous area is bordered by the Tongliang Shan, which extends roughly in a W-E direction, with peaks about 1700–1800 m above sea level. It serves as the watershed of the north-flowing tributaries of the Fen He and the south-flowing Zhang He system. The southeastern Shanxi mountains become lower southwards, merging into hilly landscapes with typical elevations of 1200–1300 m above sea level. The Zhuozhang River flows south in the central part between the two above-mentioned mountain ranges, with its height being about 990 m (around Yushe County) and 930 m (around Wuxiang County) above sea level. The presence of an upland south of Wuxiang makes the Zhuozhang River turn eastwards, then cut through the Taihang Shan, and finally empty into the Da Yunhe (the Grand Canal) on the North China Plain.

Tectonically, southeastern Shanxi is a small component of the huge Sino-Korean paraplatform. Its middle part is composed of the NNE-SSW stretching Wuxiang-Yangcheng Synclinorium (also known as Qinshui Synclinorium),

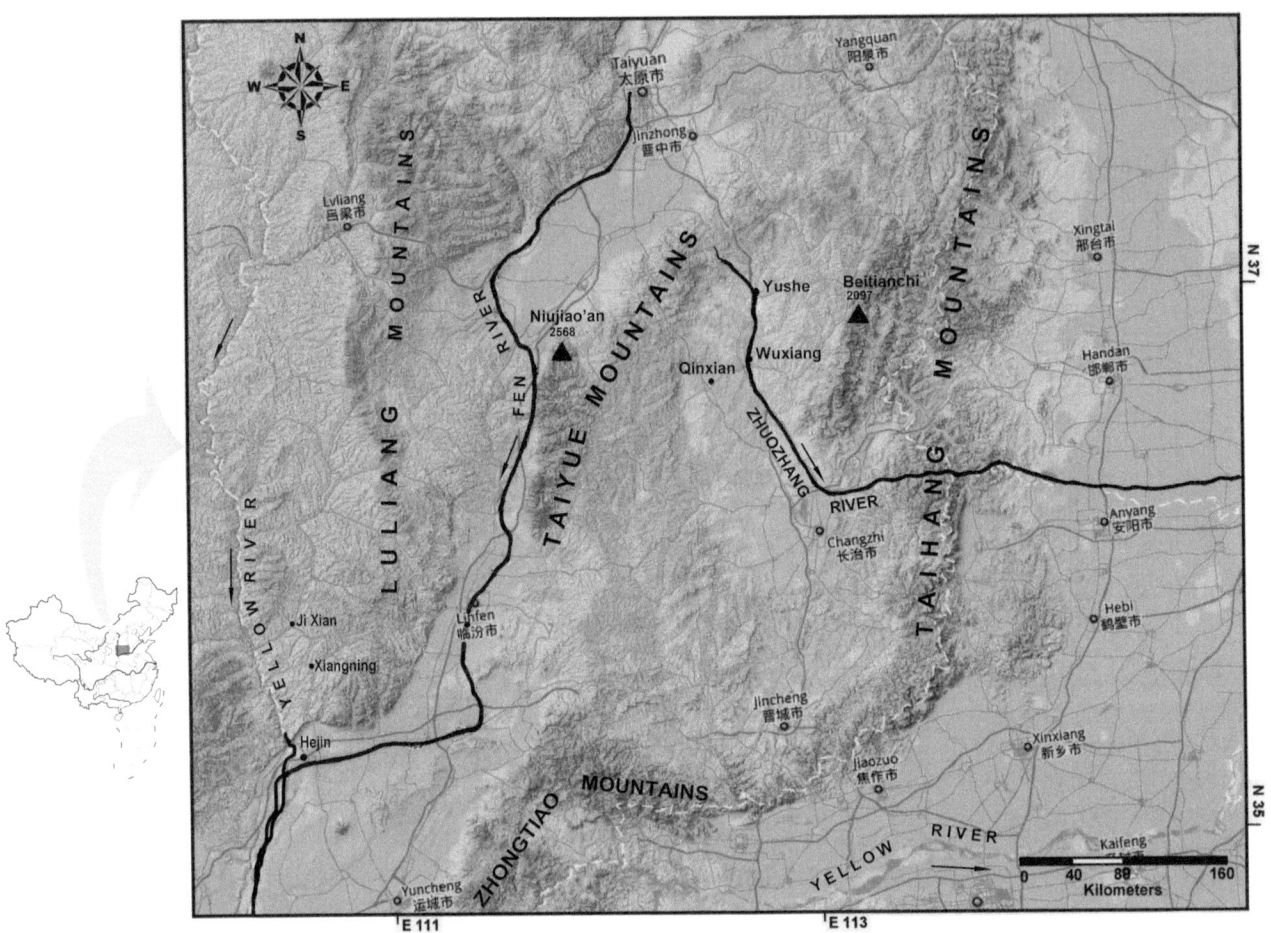

Fig. 2.1 Topographic map of southern part of Shanxi Province, south of the provincial capital, Taiyuan. Yushe is on the eastern edge of the Loess Plateau, west of the Taihang Mountains. The Yellow River (Huang He) is on the western margin of the map

bordered on the east and west sides by the Taihang Shan and Taiyue Shan. The axial part of the synclinorium is composed of Triassic red beds, which form vast hill outcrops in the axial part of the synclinorium. Fluvio-lacustrine sediments of the Yushe type are developed in a string of small basins, often of irregular form, in this axial part of the Triassic red beds. These basins can be separated roughly (from NNE to SSW) into Yushe, Wuxiang and Qin Xian basins. The sedimentary fill of the Yushe Basin is the best exposed, the most fossiliferous (especially rich in mammalian fossils), and the most extensively studied. Therefore, Yushe Basin was chosen as the target area of the present study. A number of small basins with later Cenozoic deposits, like the Siting, Xiangyuan, Tunliu, and others, are also developed south and southeast of this area (Fig. 2.2). However, they contain only incomplete portions of the Yushe Group, are not always fossiliferous, and thus are poorly investigated and paleontologically dated. They will not be dealt with in the present volume.

The Yushe Basin is not really basin-shaped in topography. If the 1100 m contour line is taken as the border of the basin, it is dendritic in configuration. The form and structure

of the Yushe Group outcrops clearly indicate their origin in an integrated fluviatile system. The basin is partitioned by several blocks of Triassic bedrock trending mainly in the ESE direction, which control interfluves of this fluvial landscape. Fluvio-lacustrine sediments occupy more than half of the territory of the present-day Yushe County and a small part of neighboring Wuxiang County to the south. A fault may exist along the Zhuozhang River as evidenced by the reappearance of Triassic bedrock along the west bank of the river. Together with the Triassic red beds, the lowest part of the Yushe Group (the Mahui Formation) reappears along the west side of the inferred fault. Therefore, the west block is the uplifted one. The alluvial deposits of the recent Zhuozhang River sometimes make correlation of the sediments on both sides of the river difficult. All this compelled us to subdivide the Yushe Basin into five subbasins: Ouniwa, Nihe, Tancun, Yuncu and Zhangcun subbasins (Fig. 2.3).

The Ouniwa and Nihe subbasins are situated in the north, the former to the northeast, and the latter west of it. They are separated from each other by a narrow strip of the uplifted Triassic bedrock along the west bank of the Zhuozhang River

Fig. 2.2 Sketch map showing the location of the Yushe Basin, north of the Wuxiang Basin in the Qinshui Synclinorium (*heavy black curve*), and locations of other Late Neogene basins of Shanxi (excerpt from *Geological Map of Shanxi*, Cui and Wu 2002)

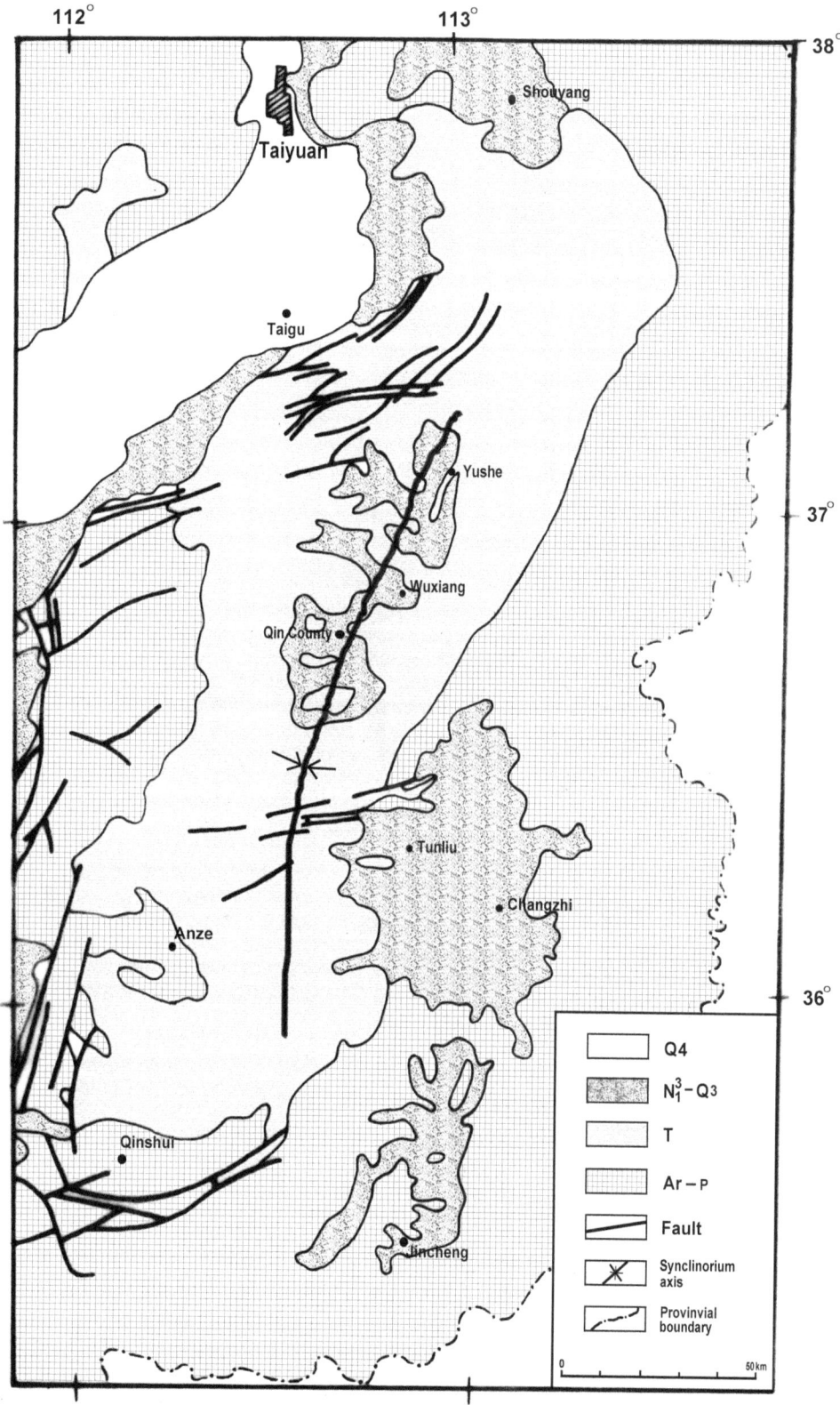

and recent alluvial deposits. The Nihe Subbasin is separated from the Yuncu Subbasin by a ridge of uplifted Triassic bedrock. However, the separation of the Ouniwa subbasin from the Tancun is obscured by the alluvial deposits developed at the border of the two subbasins. The middle pair, the Tancun and Yuncu subbasins, are clearly separated from each other by the

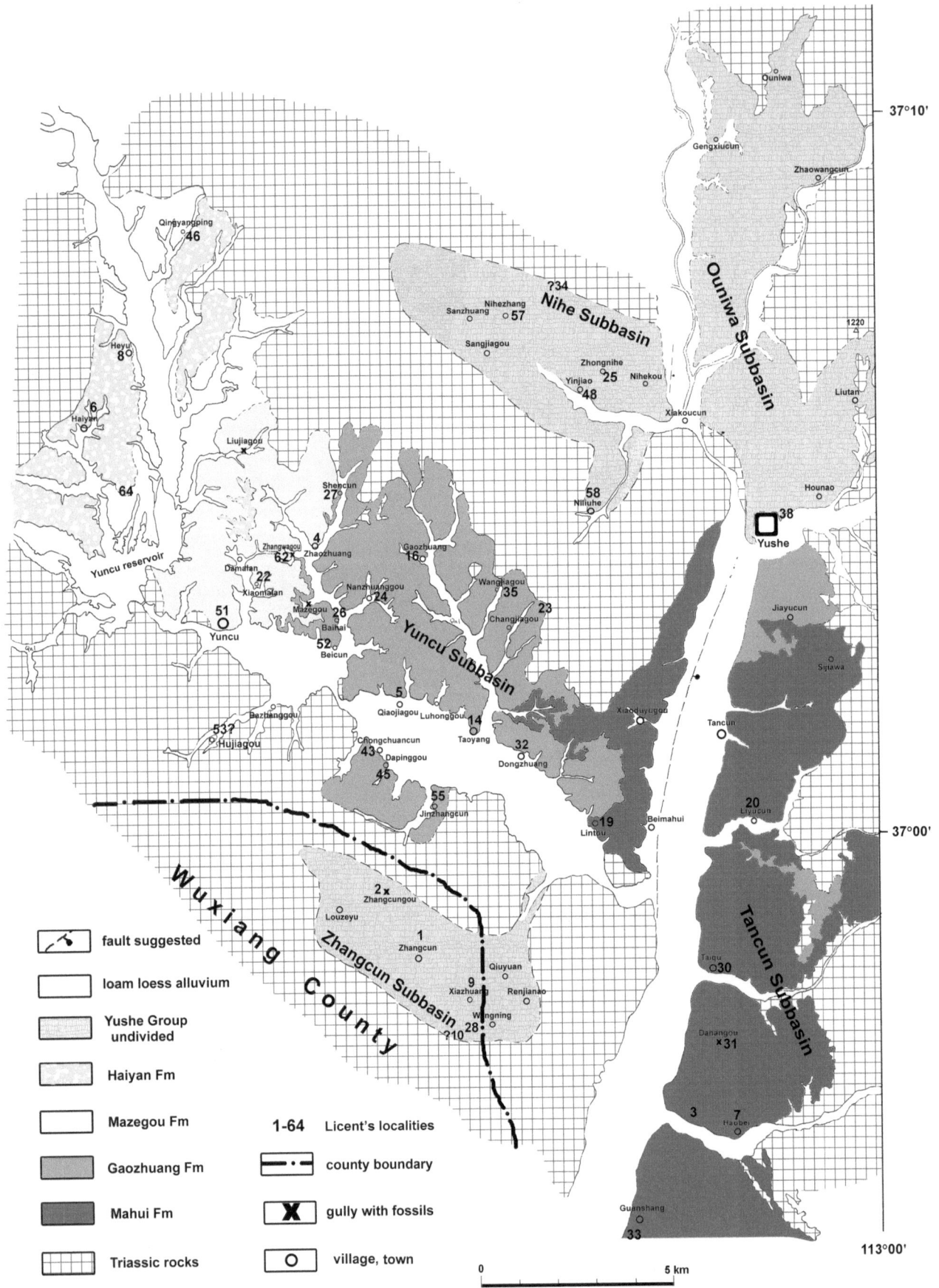

Fig. 2.3 Sketch map showing the distribution of Tertiary sediments in the five subbasins of the Yushe Basin (*numbers* denote Licent's localities)

uplifted Triassic bedrock and the alluvial deposits of the Zhuozhang River. The Tancun subbasin takes the form of a long strip, extending far south of the level of the southern-most Zhangcun subbasin. The Yuncu Subbasin is the largest sub-basin, stretching farther westward than the other subbasins. To the south the Yuncu subbasin is separated from the southern-most Zhangcun subbasin by Triassic bedrock, with a narrow connection at their east ends. The major part of the Zhangcun subbasin lies in Wuxiang County, and is restricted to the west of the Zhuozhang River and the Guanhe reservoir.

Of the five subbasins, the Yuncu subbasin has the longest and most complete sequence of late Cenozoic deposits. The composite section measured by the SAYP (Sino-American Yushe Project) starts from the basal breccias of the Mahui Formation lying directly on Triassic bedrock near Beimahui in the southeastern corner of the subbasin. The Gaozhuang and Mazegou Formations overlie the Mahui Formation, and the latest deposits of the Yushe Group, the Haiyan Formation, occur in the westernmost part of the subbasin. The late Ceno-zoic deposits of the Ouniwa and Tancun subbasins are com-parable in lithology with the lower part of the section of the Yuncu subbasin, which is equivalent to the Mahui Formation and probably the lower part of the Gaozhuang Formation. The Nihe Subbasin contains only a small part of the Mahui For-mation and a major part of the Gaozhuang Formation. The Zhangcun subbasin contains probably the second most com-plete sequence of late Cenozoic deposits. Its basal to upper parts are comparable with the uppermost Mahui through Mazegou Formations of the Yuncu subbasin, while its middle part, the Zhangcun Formation, is mainly thin-layered varie-gated clays intercalated with paper-thin oil shale, a lithology not seen in the other subbasins. The Yuncu subbasin is not only the largest, but also the most fossiliferous among the five subbasins. The majority of the mammalian fossils in the his-toric collections came from this subbasin. Thus, it was chosen as the initial area for investigation with the aim of extending the study to the whole of the fluvio-lacustrine deposits in the other subbasins.

2.2 History of Fossil Collection and Related Geological Survey

2.2.1 Early Historical Records

It is well known that in historic times the mammal fossils of China had long been called "dragon bones" by the common people and treated as pharmaceutical material by medical scholars. Some (ex. gr., Jia and Zhen 1978) believe that the "dragon bone" had been known well before the Christian era, because this word first appeared in "Shan Hai Jing" (a book about the legendary stories of mountains and seas,

allegedly taking place in the period from the Warring States to the Western Han Dynasty [476 B.C. to 8 A.D.]), collated and published by Liu Xiang (77-6 B.C.) and his son Liu Xin (?–23 A.D.) of the Han Dynasty. A careful reading of the "Shan Hai Jing" reveals that the phrase including the word is: "There are plenty of Tian Yong [an unknown kind of plant] whose form looks like the dragon's bone" (Liu et al. 2009 edition: 136). Thus "dragon's bone" is here used as a descriptor for the plant, not necessarily referring to the fossilized materials called "dragon bones" by later Chinese pharmaceutical scholars. Needham (1959: 619) suggested that the name "dragon bone" could be traced back to 133 B.C. according to historical records. At present it can hardly be ascertained whether this "dragon bone" is the same as that used later for pharmaceutical materials.

According to Li Shizhen's "Compendium of Materia Medica" (finished in 1578, but published posthumously in 1596), the first reliable record of the "dragon bone" as a kind of pharmaceutical material was in a medical book (no longer extant) written by Lei Xue (Lei Hiao in Andersson 1934), who lived in the Song Dynasty (420–479 A.D.) of the Southern Dynasties. There Lei Xue wrote: "The dragon bones from Shan Zhou [now Sheng Xian area, Zhejiang Province], Cang Zhou [now Cangzhou area, Hebei Province], and Taiyuan [capital city of Shanxi Province] are the best…". In another book written by Tao Hongjing (456–536 A.D.) of the Liang Dynasty of the Southern Dynasties, Shanxi was listed as the only place producing "dragon bones." As now known, Baode and Yushe had been the two major areas producing "dragon bones" in the early decades of the last century in Shanxi. This is fully in accordance with the historical records. Therefore, it may be safe to say that the earliest history of "dragon bones" can be traced to the fifth century, at least for the people living in Shanxi.

2.2.2 Initial Scientific Exploration (1918–1933)

Andersson's Collecting Campaign (1918–1923)

It is well known that J. G. Andersson, a Swedish geologist, was the foremost pioneer in finding fossil concentrations and organizing in situ excavations of *Hipparion* faunas in North China in the early years of the last century (Fig. 2.4). Adventurous stories of Andersson's scientific activities in China have been related in various publications (e. g. Andersson 1919, 1922, 1923, 1934; Mateer and Lucas 1985). Andersson was engaged as a mining advisor to the Ministry of Agriculture and Commerce of the Chinese government in 1914. Stimulated by frequent discovery of fascinating mammalian fossils during his geological survey

Fig. 2.4 Johan Gunnar Andersson (1874–1960)

in North China in 1916, he soon became an ardent fossil hunter. Lacking reliable information about the provenance of the "dragon bones" (mainly mammalian fossils) from the drugstores and pharmacy markets, Andersson turned to distributing circulars among the foreign missionaries in late 1917, soliciting help in providing information about the provenance of the "dragon bones and teeth" (Andersson 1919). Andersson's initiative was rewarded by information about the localities producing rich mammal fossils, including those from southeastern Shanxi. Andersson's most productive paleontological collecting activity lasted from 1918 to 1923 (vide Sefve 1927: 5) and resulted in establishment of the famous "Lagrelius Collection" at Uppsala University, Sweden.

It is a pity that no formal record of the first mammalian fossils obtained by Andersson's collectors from Yushe County can be found. Andersson did not pay much attention to the localities that yielded only poor fossils discovered during his early exploration. He may not have visited the Yushe-Wuxiang area as he never mentioned the county

name Yushe in his publications. His emphasis first focused on localities along the Yellow River (Huang He) banks of Henan Province in 1918, and later on the localities around Baode County in northwestern Shanxi in 1919–1922.

A perusal of literature and old documents revealed that the first mammalian fossils from Yushe and the adjoining Wuxiang counties were found by Andersson's collectors Liu (Chang-Shan Liu) and *Pai* (properly, Bai), probably in the second half of 1922. Andersson sent his collectors to southern Shanxi as early as 1919, but the first find was made not in the Yushe-Wuxiang area, but around the Hejin district, according to the information provided by A. B. Lewis of the Protestant China Inland Mission at Hejin, a town in southern Shanxi. A small part of the Lagrelius Collection left in China is now housed at the Institute of Vertebrate Paleontology and Paleoanthropology (IVPP). On tags stuck to a lower jaw and a maxilla of *Equus* sp. we found the following words: "Lok. 32, *Chi Hsian* [Ji Xian County, 55 km north of Hejin], Liu, 19, 12, 1919." Judging from the locality numerical sequence, the localities of the *Hsiang-ning* (Xiangning, 40 km north of Hejin) County (Lok. 33–34) were probably discovered at the same time. Both Ji Xian and Xiangning counties lie immediately north of Hejin City (see Fig. 1.1). On tags stuck to three "*Cervavitus*" jaws we read: "*Shansi* [Shanxi], *Yü She Hsien* [Yushe County], S 15 li, Tancun, NE 3 li, Chu Tse Wa [this small town does not appear on modern maps], Liu and *Pai* [Bai], 20/11, and 24/11, 1922." Probably because of the paucity of the material, the specimens found from Yushe were not given catalogue numbers. The specimens found in the Wuxiang area bear in most cases locality numbers (Lok. 70–75, 77–81). A large number of cervine jaws and limb bones, bearing the number 73 and 76 are kept in the IVPP collection. Unfortunately, no labels with year of collection have been found with these specimens. In their paper about the Lagrelius Collection, Mateer and Lucas (1985: 11) wrote the following about Andersson's 1922 activity: "*Pai* [Bai], another of Andersson's Chinese assistants, was at that time [September to December, when Zdansky was working in Baode] in southeastern Shanxi collecting a forest *Hipparion* fauna quite unlike the *Baodezhao* [Baodezhou = Baode County] fauna." This is in full accordance with the nature of the fauna from the Yushe-Wuxiang area and the locality number sequence of the Wuxiang County. Andersson's Lok. 60–63 numbers were given to the localities found during Andersson's trip to Wanping (now in Beijing) and *Chengteh* (Chengde) of *Jeho* (now Hebei Province) in May 1922 (Andersson 1923: 84, 130). All this leads us to conclude that the specimens of the Yushe-Wuxiang area must have been found by Andersson's collectors in the later half of 1922.

The material obtained by Andersson from the Yushe-Wuxiang area (see localities on Fig. 1.2), although not systematically studied until now, is by no means unimportant. A browsing of the monographs written on the Lagrelius Collection gives the following picture. Ten forms were described from Yushe Basin, as follows:

> *Canis* sp. (Zdansky, 1927), *Metailurus minor* (Zdansky, 1927), *Pentalophodon sinensis* (Hopwood, 1935), *Hipparion tylodus* (Sefve, 1927), *Hipparion parvum* (Sefve, 1927), *Dicerorhinus orientalis* (Ringström, 1927), *Rhinoceros* aff. *brancoi* (Ringström, 1927), *Propotamochoerus hyotherioides* (Pearson, 1928), *Procapreolus latifrons* (Zdansky, 1925), and *Gazella gaudryi* (Bohlin, 1935).

Among these fossils, specimens representing the two *Hipparion* species are holotypes, and that of *P. sinensis* was later chosen as the holotype of *Anancus sinensis* by Tobien et al. (1988).

The material collected from the Wuxiang area is more abundant than that of Yushe. About 20 forms were described or listed in the literature. The most important are the skulls of *Hipparion platyodus* (type, Lok. 70: Xigou, Wuxiang; Sefve, 1927) and *Hipparion ptychodus* (type, Lok. 73: Doujiaogou, Dongcun, Wuxiang; Sefve, 1927), partial skeleton of *Chleuastochoerus stehlini* (Pearson, 1928), skulls, jaws and postcranial skeletons of *Honanotherium schlosseri* (Bohlin, 1926), antlers of *Eostyloceros blainvillei* (type, Lok. 81: Hejiannao, Wuxiang; Zdansky, 1925), and large numbers of antlers, jaws and postcranial bones of *Cervocerus novorossiae* (Zdansky, 1925). Unfortunately, stratigraphic observations on these fossiliferous deposits were never reported by Andersson. Only later in 1933 did Teilhard de Chardin and Young present two sections (their Figs. 9, 10) showing the Cenozoic sediments between *Wuhsianghsien* [Wuxiang County] and *Chinhsien* [Qin Xian County] to be "fundamentally Pontian," but "overlain by a younger Sanmenian series of the same exact facies."

Teilhard de Chardin and C. C. Young's Survey of 1932

In 1931 Liu Xi-Gu (*Liushikou*, in Teilhard de Chardin and Young 1933), a collector of the Cenozoic Research Laboratory of the Geological Survey of China, was sent to southeastern Shanxi to collect "dragon bones." Liu visited the Yushe and Qin Xian areas and reported the extensive presence of "dragon bones" there. This prompted Pierre Teilhard de Chardin and C. C. Young (Figs. 2.5, 2.6) to undertake a mule-back ride through southern Shanxi in July 1932. Teilhard de Chardin and Young published a short report on their reconnaissance trip in 1933, and Young's diary covering this trip was also published posthumously

Fig. 2.5 P. Teilhard de Chardin (1881–1955)

(Yang 2009). As calculated from the itinerary described in Young's diary, Teilhard, Young and Liu arrived at Shouyang railway station on July 9, 1932. With six hired mules they started their journey on July 11. Altogether they spent 18 days, from July 11 to July 27, to traverse southern Shanxi, ending at Houma City in the southwest part of the province, where they took a long-distance bus to start their way back to Beijing.

In order to see as many geological sections as possible, they first headed southward through the central part of the Shouyang basin (see Fig. 1.1), stayed overnight in Xiaxiang village (about 20 km south of Shouyang), then turned

Fig. 2.6 Yang Zhong-Jian (=C. C. Young, 1897–1979, *right*) and Zhou Ming-Zhen (=Minchen Chow, 1918–1996, *left*), photo taken in 1963

westward to Taigu County, and continued southeastward. On July 17, after having experienced their first violent rainstorm, they arrived at the town of Houmu (now renamed Gengxiu), a village in northern Yushe Basin (Fig. 2.7).

Teilhard de Chardin and Young stayed 1 day in Houmu, collected some fragmentary fossils in a section near the village, and purchased some better preserved specimens from the villagers in the evening. The purchased material included *Hipparion*, giraffid, and mastodont teeth. Their geologic observations were published in 1933, while the description of the fossils was included in Young's monograph on the mammalian fossils from Shanxi and Henan published in 1935. According to Teilhard de Chardin and Young (1933), the section at Houmu (Fig. 2.7) included three units: (1) the "torrential-lacustrine series" consisting of basal conglomerate of rounded Triassic boulders, violet sands and clays and yellow sands (layers 1–6), (2) the older "Red loam" and the younger "red loam," lying unconformably on a deeply dissected surface cut into all older rocks, and (3) the loess (Fig. 2.8). Preliminary determination of the mammals collected from the layer 2 of the section at Houmu (Loc. 26 of CRL) included *Mastodon* sp., *Stegodon* sp., *Chilotherium* sp., *Hipparion* sp., Suidae indet., *Moschus* sp., *Cervocerus* sp., Giraffidae indet. (*Alcicephalus*), *Gazella* and *Tragocerus*(?). Thus the layers 1–2 were considered "Pontian" in age (now Late Miocene). The fossils found on the surface of layer 6 were identified by C. C. Young as *"Siphneus"* and *Hyaena* cf. *variabilis*. In

1933 the presence of a true *Siphneus* (a zokor, which is an advanced, but not yet rootless myospalacine rodent) in layer 6 led Teilhard de Chardin and Young to refer the layers 3–6 to Sanmenian age (now known to be Late Pliocene).

Later in 1935, the fossils of the layer 2 were re-identified by Young as *Chilotherium* or *Rhinoceros* indet., *Hipparion* sp., *Moschus* sp., *Cervocerus* [should be *Eostyloceros*] cf. *blainvillei* (uncertain from Wuxiang, or Loc 26?), ?*Procapreolus rutimeyeri*, *Cervus* (*Axis*) *speciosus*, *Palaeotragus* cf. *coelophrys*, ?*Microtragus* sp., *Mastodon borsoni*, *Stegodon* sp. nov. (to be described in a later volume of this series), and *Siphneus zdanskyi*. Those of the layer 6 were re-identified as *Siphneus* [now *Allosiphneus*] cf. *arvicolinus* and *Hyaena* cf. *variabilis*.

Of these fossils the mandible of *Hyaena* cf. *variabilis* is still kept in the IVPP Collection (V 9800), and has been identified as *Chasmaporthetes* sp. by Qiu and Tedford, and discussed in the chapter on Hyaenidae of Volume III of this Springer series (under preparation). This would indicate that the uppermost part of "layer 6" is of Gaozhuangian age. A well-preserved suid lower jaw (RV 3501) called *Microstonyx erymanthius* and purchased from Tancun (questionable locality) is now identified as *Microstonyx major sinensis* by Li (to be published in Yushe Volume V).

At noon of July 18, Teilhard de Chardin and Young arrived at Tancun, a "dragon bone" purchase and transportation center. As stated by Young, the village head was the dominant "dragon bone" dealer. In his yard Teilhard de

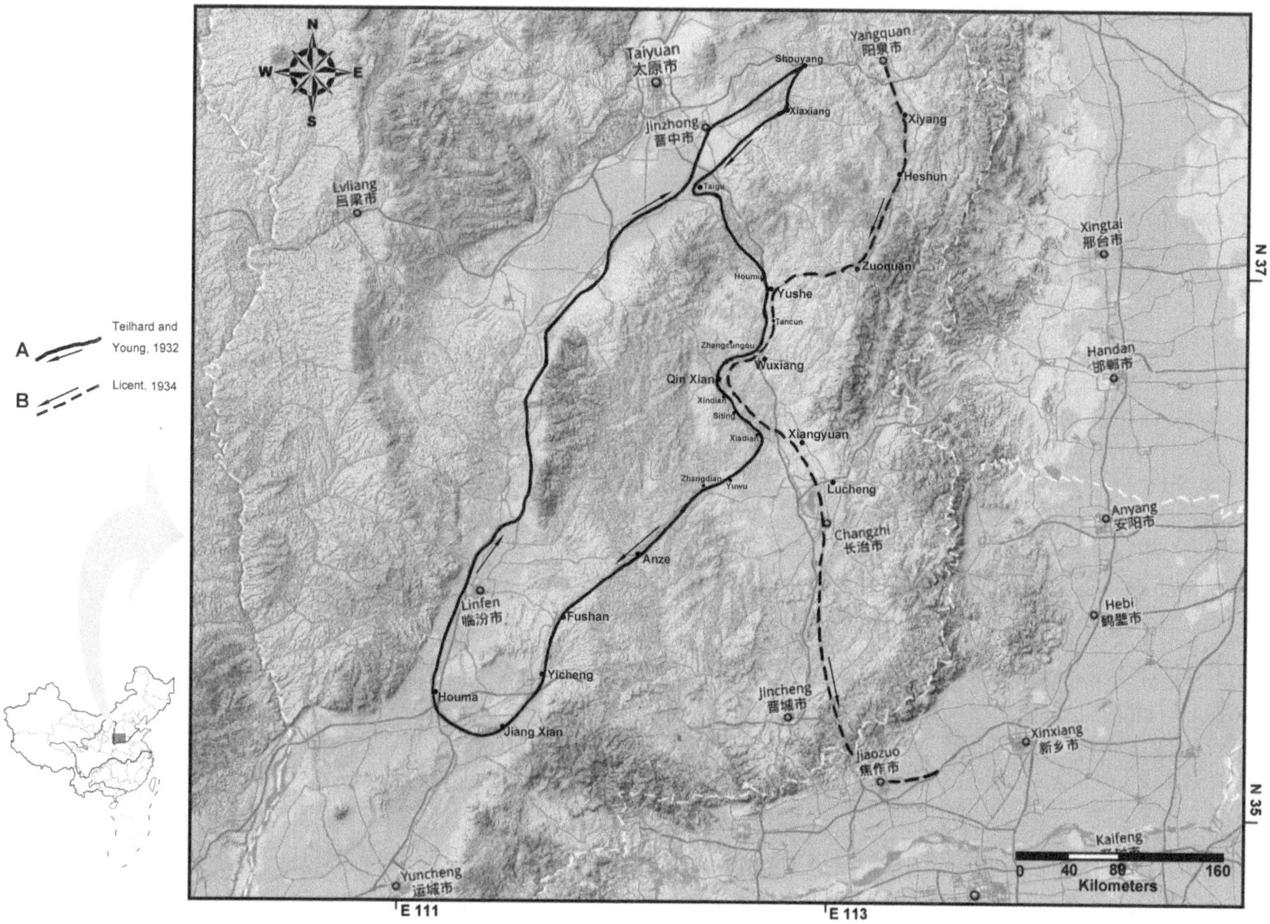

Fig. 2.7 Map showing geological reconnaissance routes in southern Shanxi followed by Teilhard de Chardin and Young, and later by Licent. *A* (*solid line*) Teilhard de Chardin and Young, 1932; *B* (*dashed line*) Licent, 1934

Fig. 2.8 Section near Gengxiu (originally Houmu, in Ouniwa Sub-basin), drawn by Teilhard de Chardin and Young 1933. *Tr* Triassic beds; *V* violet freshwater series: *1* basal conglomerate, *2* lower sands, *3* dark, plant-bearing clays, *4* middle sands (turtle sands), *5* yellow sands and limestones, *6* upper sands; *R* red loess (originally "Red Loem"); *L* younger loess (originally considered the "Loess")

Chardin and Young saw "heaps of sadly broken bones," and in one of his rooms saw complete machairodont and gazelle skulls, suid jaws, proboscidean teeth, limb bones of *Hipparion*, and broken deer antlers, etc. The village head admitted that he first bought the fossils from nearby villages, and that they were then sold at Qin Xian, an intermediate county seat of the Changzhi-Taigu motor road, where the fossils were to be transported by buses and train to Beijing, or Tianjin. The village head even said that the better traffic service made the "dragon bone" trade in Yushe a more advantageous position

than that in Baode. At the end of their visit, after much hard bargaining, Teilhard de Chardin and Young purchased five or six of the best preserved specimens at still very high prices, for example, one and a half dollars for a suid lower jaw.

Teilhard de Chardin and Young next visited *Changchiakou* (Zhangcungou) in Wuxiang County. Then they went first southwestward to *Chinhsian* (Qin Xian County), and then southeastward to *Hsintien* (Xindian, some 20 km [not 30 km as stated by Teilhard de Chardin and Young] from Qin Xian County seat), where the "Pontian" freshwater series ended.

Four hand-sketched profiles of the sections near *Chang-chiakou* and *Chinhsian* were published (Teilhard de Chardin and Young 1933, their Figs. 8–11). Their Figs. 9 and 10 showed clearly that the Cenozoic sediments between Wuxiang and Qin Xian contained "Pontian" faunas overlain by similar deposits with "Sanmenian" faunas. Confined to observations adjacent to the motor-road, they did not get a complete impression of the lateral extent of the basin fill, but the general stratigraphy and setting were accurately assessed. The late Cenozoic deposits were seen to be contained within the axial part of a synclinorium developed in Triassic rocks and to dip westward more or less in conformation with the eastern limb of the Triassic fold, so that the basal parts of the late Cenozoic strata were exposed only along the eastern borders of the basins.

Finished with the Yushe-Wuxiang-Qin Xian areas, Teilhard de Chardin and Young turned southwestward, passing through *Yüwuchen* (Yuwu town) and Zhangdian of Tunliu County, then to *Fuchengchen* (Fucheng town, now Anze County), Fushan, *Icheng* (Yicheng County), and *Chianghsien* (Jiang Xian County), and finally, on July 27, arrived at Houma city. From Houma city Teilhard de Chardin and Young returned to Beijing.

2.2.3 *Period of Extensive Collection and Study (1934–1945)*

Licent's Collecting Campaign (1934–1935)

The discoveries made by Teilhard de Chardin and Young were thought of particular importance at that time because the fluvio-lacustrine character of the "*Hipparion* fauna"-bearing deposits radically differed from the classical *Hipparion* red clay facies widely distributed in North China. This fact aroused the interest of the French Jesuit Father Emile Licent of the *Musée Hoang-ho Pai-ho de Tientsin* (now Tianjin Natural History Museum).

Emile Licent (Fig. 2.9) was an important figure in the history of the development of the natural history museums in China. As a missionary he was sent to China as early as 1914. However, his main goal was apparently to initiate natural science investigations in China. Early on, he had already obtained large paleontological collections from Qingyang (Late Miocene *Hipparion* fauna) in Gansu, *Sjara-osso-gol* (now Hongliuhe, Wushen Banner; Late Pleistocene fauna) in Nei Mongol, and Nihewan (Early Pleistocene *Equus* fauna) in Hebei. He eventually launched a new campaign of collecting fossils in the Yushe area in the mid-1930s, resulting in obtaining about 2,300 specimens there. Licent left China in 1938, 1 year after the Anti-Japanese War broke out and Tianjin was occupied by Japanese troops.

Fig. 2.9 Emile Licent (1876–1952, *left*) in the 1930s at the *Musée Hoang-ho Pai-ho de Tientsin*

Licent left Beijing on June 8, 1934. Arriving at Shijiazhuang (capital city of Hebei Province) on June 9, he first turned westward to Yangquan, then turned southward along the west side of the Taihang Shan Mountains, via Xiyang, Heshun and Zuoquan (then Liao Xian), then turned westward again, and on June 26 he arrived at Lintou village of the Yuncu Subbasin (Figs. 2.7, 2.10), where there was a Catholic parish church, then led by a Dutch Father, P. Landolinus Bonekamp.

Licent started his prospecting and excavation first at a gully east of Lintou, called Heilingou (differently transliterated as *Hu Lien Keou, Ho Lien Keou,* and *Ho Lin Kou* in Licent's diary, vide Appendix III). The work at Lintou had to be terminated on July 3 because of the unaffordable demand for compensation asked by the landowner, according to our interviews in 1994 and 1997 with the villagers who knew of Licent's activities in those days either directly or indirectly from their older generations. Licent then went to Zhangcungou (*Tchang tsoun keou* in Licent's diary) in the Zhangcun Subbasin. He carried out major field work there and stayed until July 26. On July 27 he returned to the Yuncu Subbasin, stayed in Shencun (*Chen ts'ouna* in Licent's diary) until August 2. Licent collected and purchased fossils widely in the area nearby

Fig. 2.10 Church in Lintou village (built in 1884; photo by Z.-X. Qiu, August 1997)

from Gaozhuang (*Kau tchoang* in Licent's diary) in the east, at Baihai (*Pai hai tze* in Licent's diary) in the south, and from Damalan in the southwest and Haiyan (*Haiyen* in Licent's diary) to the west of Shencun. On August 3 he returned to Lintou to wind up field work. He stayed there until August 10. From his diary it is clear that Licent not only organized excavations in the areas near Lintou, Zhangcungou and Shencun, but also extensively purchased specimens from villagers living nearby at the time. On August 10 Licent left Lintou for a reconnaissance trip in southeastern Shanxi. He prospected large areas of Qin Xian, Siting, Lucheng, Changzhi, etc. Then, via Jiaozuo of Henan Province, he headed northward and was back in Tianjin on September 26, 1934.

According to Licent's diary and recollections of the local villagers, Licent was accompanied by two Chinese facilitators. The villagers of Lintou told us that they were called Mr. Zhang and Mr. Wang, but the villagers of Zhangcungou insisted they were bodyguards named Wang and Tian. One of the "dragon bone" collectors hired by Licent was "*Lu cull*," a name which appeared very often in Licent's diary as a major helper and collector. A third Chinese facilitator called "*Lei hoie tchang*," sometimes "*Lien tchang*," was a

local "dragon-bone" dealer, whose Chinese name should be correctly spelled as Hao Lin-Zhong according to the opinion of the Zhangcungou villagers. Many old villagers of Zhangcungou still remember him, saying that he was a native of Shibi, a little town about 5 km southeast of Zhangcungou, but belonging to Wuxiang County. He served as an agent dealing with the local people on behalf of Licent. Another local Chinese who played a rather important role in excavation near Zhangcungou was "old Ho," a villager of Zhangcungou. He directed most of the excavation work there. Father M. Trassaert, an entomologist, who joined the *Musée Hoang-ho Pai-ho de Tientsin* in 1934, came to Zhangcungou on July 5. He participated mainly in the work of the Zhangcungou area. Licent always hired local people as laborers, sometimes more than ten people, plus as many as seven mules or donkeys.

Licent visited the Yushe area again in 1935, first departing from Tianjin on May 22. This time he continued westward beyond Shijiazhuang to Yuci, then turned southward and arrived at Yuncu on May 29. He stayed mainly in the western part of the area, visiting Damalan, Zhaozhuang, Haiyan, Gaozhuang, and elsewhere. He did not organize any excavation this time, but purchased many specimens during his travel. On June 9 he stayed overnight in Lintou and left for Wuxiang the next day. From there he went to Changzhi and then turned westward to prospect areas in Huoshan County. On June 18 he returned to Yuci via Taigu, and arrived at Tianjin on June 19.

After their first year's work, Licent and Trassaert published a short article in 1935. In this article the studied area was considered to comprise three "minor basins": *Changtsun* (Zhangcun), Yushe, and *Yünchuzhen* (Yuncu). The Yuncu basin was the largest and was selected as the most representative of the three. The smaller Zhangcun basin was characterized by the particular lithology of the middle part of the deposits (thick "marlish beds," with abundant fish remains).

The lithology of the Yushe basin was not indicated in their geologic sketch-map. The "Pliocene lacustrine deposits" were subdivided into three "zones," based on lithology and paleontology. All the fossil mammals listed in their article were preliminarily identified by Teilhard de Chardin, who claimed partial authorship and made a number of corrections to reprints of that article (analysis of Schmitz-Moormann 1971, Tome V: 2232, 2241–2242).

Zone 1 was "mostly exposed and fossiliferous near *Lingt'ou* [Lintou]" and represented by "hard consolidated conglomerates and dark red sandstone, immediately derived from the underlying Permo-Triassic beds." "It contains a typical Pontian fauna" (Licent and Trassaert 1935).

Zone 2 was best studied in the Zhangcun subbasin. Deposits were "less coarse, and a typical lacustrine condition is prevailing: green and bluish marls, containing many bird, turtle, fish-remains, freshwater shells… and plant remains." A "remarkable type of strepsiceros Antilope" [cf. *Antilospira*] and a specimen of "*Castor majori*" [later transferred to *Dipoides*] "characteristic of the Ertemte fauna" were collected. This led the two French authors to conclude that "a middle Pliocene age seems to be indicated" by the mammalian fauna (Licent and Trassaert 1935).

Zone 3, studied in the Yuncu Subbasin, was "chiefly sandy, with the presence however of two layers of marl at the middle of the deposits … indicating perhaps a maximum in the lacustrine conditions" (Licent and Trassaert 1935). Zone 3 rests with "clear erosional breaks and even overlapping" upon Zone 2. As to the age of Zone 3, Licent and Trassaert were inclined to separate it into two phases: the lower part containing "a big *Hipparion*, already associated with *Bison*, but without any sure trace of *Equus*," and a "thick lamelled Elephant," which should be older than the upper part containing the typical *Equus* fauna.

Licent and Trassaert's work was of considerable importance, for it remained, for 40 years, the basis for the geology and stratigraphy of the Yushe Basin fill. Nevertheless, a series of ambiguities and inconsistencies exist.

Firstly, the geologic sketch-map appended to the article (Fig. 2.11 here) showed basically concentric patterns of distribution of the deposits, while in the text it was clearly stated that the deposits were all westward dipping. Furthermore, the largest area of outcrops as shown in that map was that of the Zone 3. This is far from being true as proven by our own observations during the 1987–1988 field seasons.

Secondly, the composite section covering the deposits of the three zones taken from a cliff near Zhangcungou (Fig. 2.12, left) seemed rather misleading in that Zone 2 included also the lower half of the red loam, and the upper half of the red loam represented Zone 3. Furthermore, the dark red sandstone of Zone 1 should not have appeared in this area if the deposits were all westward dipping. Huang and colleagues measured the same section, but referred these deposits to layers 26–34 (upper member) of the Zhangcun Formation (Huang and Guo 1991, their Figs. 1–6; here Fig. 2.12, right). Our re-examination of the Zhangcungou cliff section in 1997 revealed that the deposits exposed at the very base of the cliff are still yellow sandstones, and a left mandibular ramus of *Paramachairodus schlosseri* was found there in situ (vide Chapter of the Felidae, Volume III of the present series), and we purchased some cheek teeth of *Hipparion houfenense* (vide Chapter of the Equidae of volume IV of the present series). The base of the section would therefore belong to Zone 2. The faunal lists given by Licent and Trassaert (1935) for their Zones 1–3 seemed to be based largely on the evolutionary levels of the listed mammals, without reliable ties with their factual stratigraphic positions in the section.

Teilhard de Chardin's Study of the Fossils Collected by Licent

During 1937–1945 the Yushe and Wuxiang region became part of the battle-zone, prohibiting field work there. However, study of the fossils collected from Yushe was carried on successfully in the laboratory by Teilhard, to whom the task was apparently entrusted by Licent. Unfortunately, Teilhard succeeded only in accomplishing half of his original plan to publish eight monographs covering all the Yushe mammalian fossils. Together with Trassaert, he published three monographs on the Yushe ungulates: on the Proboscidea, Camelidae, Giraffidae and Cervidae (Teilhard de Chardin and Trassaert 1937a, b) and on Cavicornia (Teilhard de Chardin and Trassaert 1938). With the worsening situation in Tianjin caused by its occupation by the Japanese troops, Teilhard had to transfer some of the Yushe fossils, mainly small to medium sized rodents and carnivores, to Beijing. Together with Pierre Leroy, Teilhard established a new institution in Beijing, the Institut de Géobiologie. He used this as a vehicle to publish the Yushe

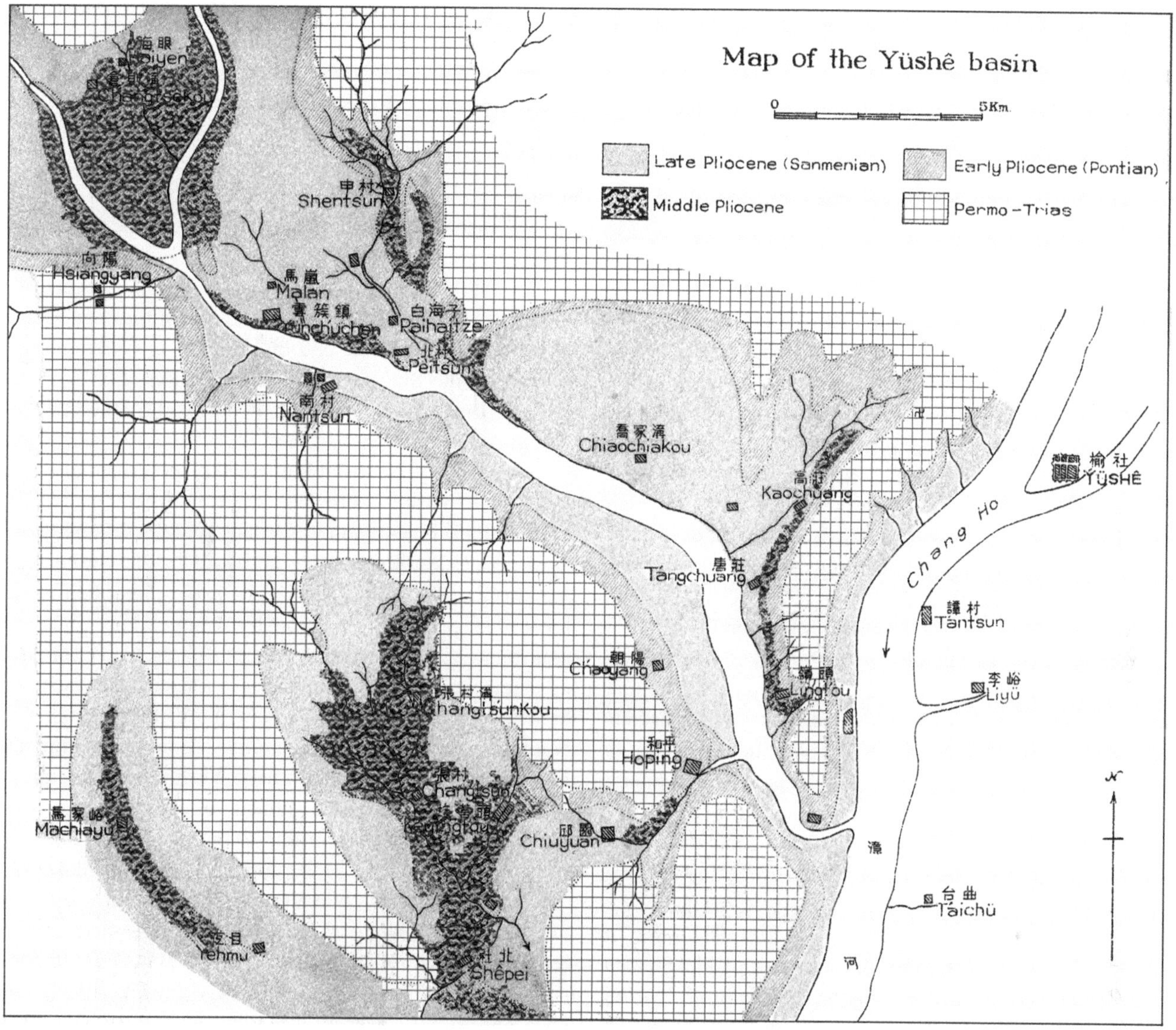

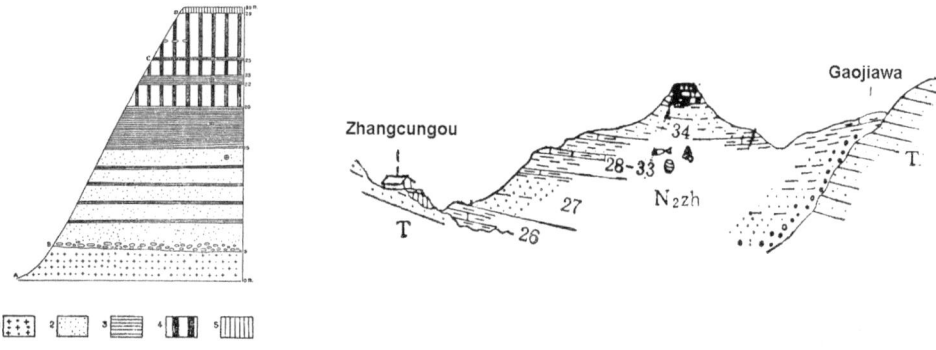

Fig. 2.11 Geological sketch map of Yushe Basin (mainly Yuncu subbasin) drawn by Licent and Trassaert (1935)

Fig. 2.12 Comparison of two sections across the excavated cliff near Zhangcungou village. *Left* section measured by Licent and Trassaert (1935): *1* dark red sandstone, *2* yellow sand, *3* dark marl, *4* red sandy loam, *5* loess; *AB* Zone 1, *BC* Zone 2, *CD* Zone 3. *Right* section measured by Huang and Guo (1991): Units 26–34 equivalent to those on *left*

fossil rodents, mustelids and felids together with their living species distributed in China through 1945 (published as Teilhard de Chardin 1942; Teilhard de Chardin and Leroy 1945a, b). When Teilhard left China in 1946, the study stopped abruptly, leaving the Canidae, Ursidae, Hyaenidae, Viverridae, Rhinocerotidae, Equidae, and Suidae unstudied. Of these latter families only a small fraction was later treated (vide infra).

Fossil Collecting Supported By Childs Frick (1932–1937)

Childs Frick, curatorial associate of the American Museum of Natural History, started to support the collecting of mammal fossils in Yushe from 1932. Encouraged by Walter Granger, he hired the skilled professional collector Gan Quan-Bao (*Kan Chüan Po*, nicknamed "Buckshot"), who had been trained during the museum's Asiatic expeditions in China and Mongolia (Andrews 1932). Gan's activities were funded by Frick and supervised by Dr. Erik T. Nyström, a geologist affiliated with the Sino-Swedish Scientific Research Association based in Taiyuan, the capital city of Shanxi Province. Knowing that Nyström's assistants came across some "dragon bones" in the Shouyang area in 1928, Gan started his prospecting in this area in January 1932. Gan found two areas south of Shouyang where active "dragon bone" mining was going on. Seven sites were situated west of the Zheng-Tai railroad (from Shijiazhuang [Zhengding], capital city of Hebei Province, to Taiyuan), around "*Pai Tao Tsun*," which is 16 miles west of Shouyang. Sixteen sites were located south of the Shi-Tai (then Zheng-Tai) railroad and south of "*Chang Chia Chuang* [Zhangjiazhuang]." He worked these sites, including purchase of material from 1933 to 1937. As mentioned in earlier work by Teilhard de Chardin and Young (1933: 209–212), in the Shouyang basin the violet, "fluviatile or lacustrine series" produces "*Caprolagus* sp. [*Caprolagus brachypus* in Young, 1935, now *Sericolagus brachypus*] and *Siphneus* cf. *tingi*" [*Siphneus* cf. *chaoyatseni* in Young, 1935, now *Yangia chaoyatseni*] of "Sanmenian age." Elsewhere, in the Xiaxiang (*Shiahsiang*) basin, "some 200 meters of powerful conglomerates and violet sands" contain "typical Pontian fossils" like "*Hipparion, Chilotherium, Alcicephalus* [Giraffidae indet. in Young, 1935], etc." The Frick Collection of the 1930s verifies this generalization. The records of individual taxa, occurring together *Percrocuta* (*Adcrocuta*?) and *Eostyloceros*; *Stegodon* and *Chleuastochoerus*; *Chasmaporthetes* and *Canis* or *Bison* and *Dinofelis*, suggest a temporal range for the Shouyang

area comparable to that represented by the Yushe Group 90 km to the south.

During this interval of collaboration with the Sino-Swedish Scientific Research Association, Childs Frick also engaged Xi-Gu Liu (who had visited Yushe for the Cenozoic Research Laboratory) to collect in the Baode and Fugu areas. Consequently the Frick Collection came to include material from there as well as from the Yushe and Shouyang areas.

Meanwhile, Gan got the news that the Yushe district, 90 km south of Shouyang, produced more and better "dragon bones." By the end of 1934 Dr. E. Nyström, now Director of the Sino-Swedish Scientific Research Association and living in Beijing, got to know from Teilhard de Chadin that rich fossils had been found from the Yushe area by Licent's party. Nyström immediately gave instructions to Gan to go to Yushe. While Gan was working at Zhaozhuang in May 1935, Licent came to Yuncu on May 29 and got to know about Gan's collecting activities in the Yushe area. The two parties met on June 1. To avoid direct confrontation with Licent, Gan intentionally arranged his subsequent work there in the absence of Licent's party. Gan continued his extensive collecting in 1936. As a result, he sent more than two hundred parcels to New York. Since Frick was then interested mainly in carnivores and horned ruminants, the material collected by Gan from Yushe belonged mostly to those groups. The total number of the specimens collected by Gan is 298, according to our personal count (226 formally recorded in AMNH Archives, see Appendix VI). The existence of this collection was noted several times in the literature (Teihard de Chardin and Trassaert 1937b; Teilhard de Chardin and Leroy 1945a, b), but the collection has not been studied, nor even widely known. After 1968, when the Frick Collection was donated to the AMNH, the specimens from Yushe became accessible.

In the summer of 1936, accompanied by Charles L. Camp of the University of California, Berkeley, to search for Triassic therapsid fossils in southern Shanxi, C. C. Young visited the Yushe-Wuxiang area again (Yang 2009). This time they started their journey from the Yangquan railway station, went directly southward, via Xiyang and Heshun, to Liao Xian (now Zuoquan), then turned westward to Yushe. They again visited the same "dragon bone" dealer in Tancun, and saw a large quantity of various fossils, including very well preserved skulls. They also saw a large and impressive house, which was still being built. Apparently, the dealer had made money from the "dragon bones" since 1932. Unfortunately, Young failed to purchase any fossils because of unbelievably high prices demanded by the "dragon bone" dealer.

2.2.4 Relatively Stagnant Period (1945–1975)

C. C. Young and Dong-Sheng Liu (=Liu P. T.) described (1948) some specimens obtained from Shanghai drugstores, allegedly transported from Yushe. The material included some hyaenids, rhinoceroses, and other large mammals. The type skull with mandible of *Leecyaena lycyaenoides* was the best specimen among them, and is now kept in IVPP.

Field work in Yushe was not immediately resumed after the establishment of the People's Republic of China in 1949. In 1955, a team headed by Liu Xian-Ting, a paleoichthyologist of the Laboratory of Vertebrate Paleontology, Academia Sinica, was sent to the Yushe-Wuxiang area. The primary aim of the team was to find more material of Triassic mammal-like reptiles, the first fossils of which were discovered some 20 years previously. Finding vertebrate fossils in late Cenozoic deposits was only the secondary purpose of the team. They departed from Beijing on September 12, arrived at Taiyuan by train on the 13th, and at Taigu by car on the 14th. From Taigu they used camels as the means of transportation and spent 5 days to reach the Wuxiang County seat on September 19. Excavations were primarily focused on the dicynodont fauna from the Triassic rocks in the Louzeyu area (Wuxiang County). From October 9 to 16 the team stayed in Zhangcungou to prospect the vertebrate fossils of the late Cenozoic fluvio-lacustrine deposits. Well preserved fish fossils were excavated from the greenish and bluish marls. Unfortunately, the team failed to find any important fossil mammals. From October 21 to 25 some excavations were carried out in localities near Gaozhuang (Jingjiagou, Field No. 5541; Ya'ergou, Field No. 5543) of the Yuncu Subbasin, where a number of *Hipparion* teeth were found (vide Chapter on Equidae of Volume IV of the present series), and the type material of the murid rodent *Chardinomys yusheensis* was collected (Jacobs and Li 1982). From October 25 to November 2 excavations were successfully carried out at Hounao, a new locality 1.7 km northeast of the Yushe county center. Well preserved specimens of rhinos, suids and cervids were unearthed there. During this time a delegation of vertebrate paleontologists from the Soviet Union headed by J. A. Efremov, visited some of the excavation sites (Zhangcungou and Hounao). The delegation arrived by train at Yuci on October 27, visited Zhangcungou on October 29, Hounao on 30th, and departed Yushe on October 31. The field work ended on December 2. As far as the mammalian fossils are concerned, Hounao was the major new locality discovered in 1955. Excavation was resumed at Hounao in 1956 (Field No. 5679) with the result that the small pocket of mammalian fossils was totally exhausted.

In 1962 *Liu Hsien-T'ing* (Xian-Ting Liu) and *Su Te-Tsao* (De-Zao Su) published their description of the fish fauna

found from the Yushe-Wuxiang area in 1955–1956, and listed altogether 13 species, most of which were cyprinids (Liu and Su 1962). All these fish specimens are kept today in IVPP. Unfortunately, the fossil large mammals found from Yushe during 1955–1956 have never been systematically studied, except a few isolated *Hipparion* teeth mentioned by Qiu et al. (1987).

Chang-Kang Hu (1962) created a new species of *Metacervulus*, *M. lepidus*, based on a perfectly preserved skull donated to IVPP by a Shanghai amateur collector, allegedly found from Yushe.

In a review of the Chinese Cenozoic stratigraphy, Pei et al. (1963) first transformed Licent and Teilhard de Chardin's "Yushe series" to Yushe Formation and relegated the zones to lower, middle and upper divisions, but no real conceptual advance was made. New geological investigations were deferred until the early 1970s when the Geological Bureau of Shanxi Province began the 1:200,000 scale mapping of Shanxi Province. The Yushe Basin was contained in the Fenyang-Pingyao Quadrangle (published by the Geological Bureau of Shanxi Province in 1976). The local geologists elevated the Yushe Formation to group status, and chose the sequence in the Zhangcun subbasin as the type section for a three-fold subdivision of the Yushe Group (Renjianao, Zhangcun and Louzeyu Formations). The lithostratigraphic units as above defined owed much to concepts of the 1930s: the Rejianao Formation represented the coarse colluvium and alluvium at the base of the section; succeeded by the Zhangcun Formation, mostly fine sands and clays (including "oil shales") of a dominantly lacustrine nature; and at the top, the Louzeyu Formation sequence of sands, clays and marl representing a mostly fluviatile regime with minor lacustrine events. The latter unit was inferred to rest unconformably on the Zhangcun Formation, which truncated the Renjianao Formation itself. As measured in the Zhangcun Subbasin the Yushe Group was about 460 m thick (Fig. 2.13).

2.2.5 Renewed Study, Late 1970s and Early 1980s

In the late 1970s and the 1980s interest in studying the geology and paleontology of the Late Cenozoic deposits of the Yushe-Wuxiang area was renewed principally by three groups of workers as highlighted in the following. The emphasis of research became increasingly multidisciplinary.

Geological studies of Beijing University. During the 1970s and the first half of the 1980s, Jia-Xin Cao, her students, and coworkers of Beijing University, conducted a geologic survey of the Taigu-Yushe-Wuxiang area. In the Yushe Basin they concentrated their work in Zhangcun

The Zhangcun Subbasin

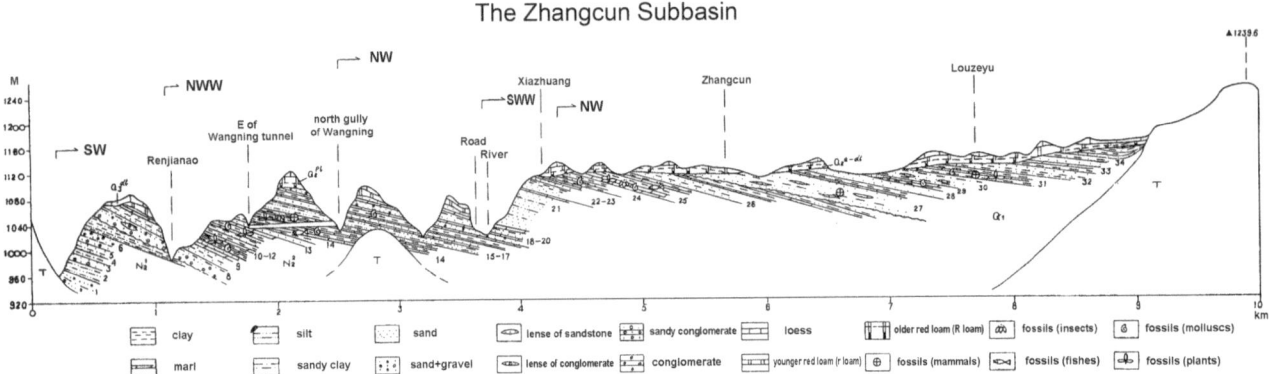

Fig. 2.13 Section of Zhangcun Subbasin measured by the geologic mapping team of the Geologic Bureau of Shanxi Province (redrawn from Plate 17 in the explanatory text of the Fenyang-Pingyao quadrangle, 1976)

subbasin (Cao 1980; Cao and Wu 1985; Cao and Cui 1989). Their study laid emphasis on reconstructing the depositional paleoenvironment, based on stratigraphic divisions proposed by the Shanxi geologists with the implication that major events in the Zhangcun subbasin held for the whole of the Yushe Basin. From the point of view of vertebrate paleontology, the most important find during their survey was a slab with "*Chilotherium*" skeletons from the basal level of the Zhangcun Formation near Wangning village in 1980, now kept in the Yushe County Museum (Fig. 2.14). The palynological record (Cao and Cui 1989) showed that the mesic floras of the lower part of the Zhangcun Formation represented a subtropical to warm-temperate climate, while the more xeric assemblages in the upper part of the Zhangcun Formation represented a warm-temperate climate approximating that of the southern part of North China of today.

A major contribution to the study of the Yushe area was made by Ning Shi, then a postgraduate student of Jia-Xin Cao, during the years of 1986–1993. In a comprehensive monograph on the Zhangcun subbasin, Shi (1994) created a new Formation, the Wangning Formation, lying between the Renjianao and Zhangcun Formations. Based on palynologic, mineralogic, paleomagnetic and [10]Be investigations, Shi was able to date the four Formations of the Yushe Group in the Zhangcun subbasin as follows: Renjianao: 5.5–4.5 Ma; Wangning: 4.5–3.5 (or 3.4) Ma; Zhangcun: 3.5 (or 3.4)–2.3 Ma and Louzeyu: 2.3–1.5 Ma. Sedimentation in that area began at about 5.5 Ma, some million years after deposition began in the Yuncu subbasin.

Paleomammalogy research by IVPP. In 1978 Zhan-Xiang Qiu, then head of the newly established Neogene Division of the IVPP, started a plan to resume the study of the Yushe fossil mammals kept in the Tianjin Natural History Museum (TNHM), and left unstudied by Teilhard de Chardin. A series of geologic surveys was conducted, with particular efforts to obtain more reliable information about the provenance of the fossils to be studied. Qiu first visited the Yushe Basin in October 1978. Later Qiu and his colleagues (Hang Jia, and Wei-Long Huang of the TNHM) started extensive prospecting of the area in 1979 (February) and 1980 (May and July–August). During these surveys mammal fossils were also collected, some in situ, and some purchased (see Appendix VII). In July–August 1985, Qiu, De-Fa Yan, Hang Jia and Wei Dong studied the Yuncu Subbasin in more detail, and measured some sections.

The field work revealed that the Yuncu subbasin might be more representative of the whole Yushe Basin than the Zhangcun subbasin. The Yuncu subbasin is almost totally separated from the Zhangcun with only a possible connection by a narrow strip of outcrops of fluviatile sediments. The Yushe Group in the Yuncu subbasin is more complex, dominated by fluviatile environments, and nearly twice as thick as in the Zhangcun subbasin. Qiu et al. (1987) resolved the lithostratigraphic sequence in the Yuncu subbasin into four sedimentary cycles, each beginning with coarse clastics and fining upward. Accordingly, a new lithostratigraphy was proposed: The Mahui Formation, containing a basal colluvium and characterized by yellow cross-bedded sands fining upward to muddy sands, clays and marls. It is overlain, locally unconformably, by the Gaozhuang Formation dominated by fluviatile sands and ending in mudstones and marls. These deposits are disconformably overlain by the Mazegou Formation, hard muddy sandstones and mudstones with its top often obscured by alluvium. The final unit, the Haiyan Formation, is horizontal and thus lies with angular unconformity across the gently dipping Mazegou Formation. It is composed mostly of fine sandstones and siltstones with local clays and marls.

Meanwhile Qiu et al. (1980, 1987) studied the hipparionine fossils of Yushe and other late Neogene basins. Qiu (1987) also published a monograph on the Pliocene and Pleistocene Hyaenidae of the Yushe area, based on his study

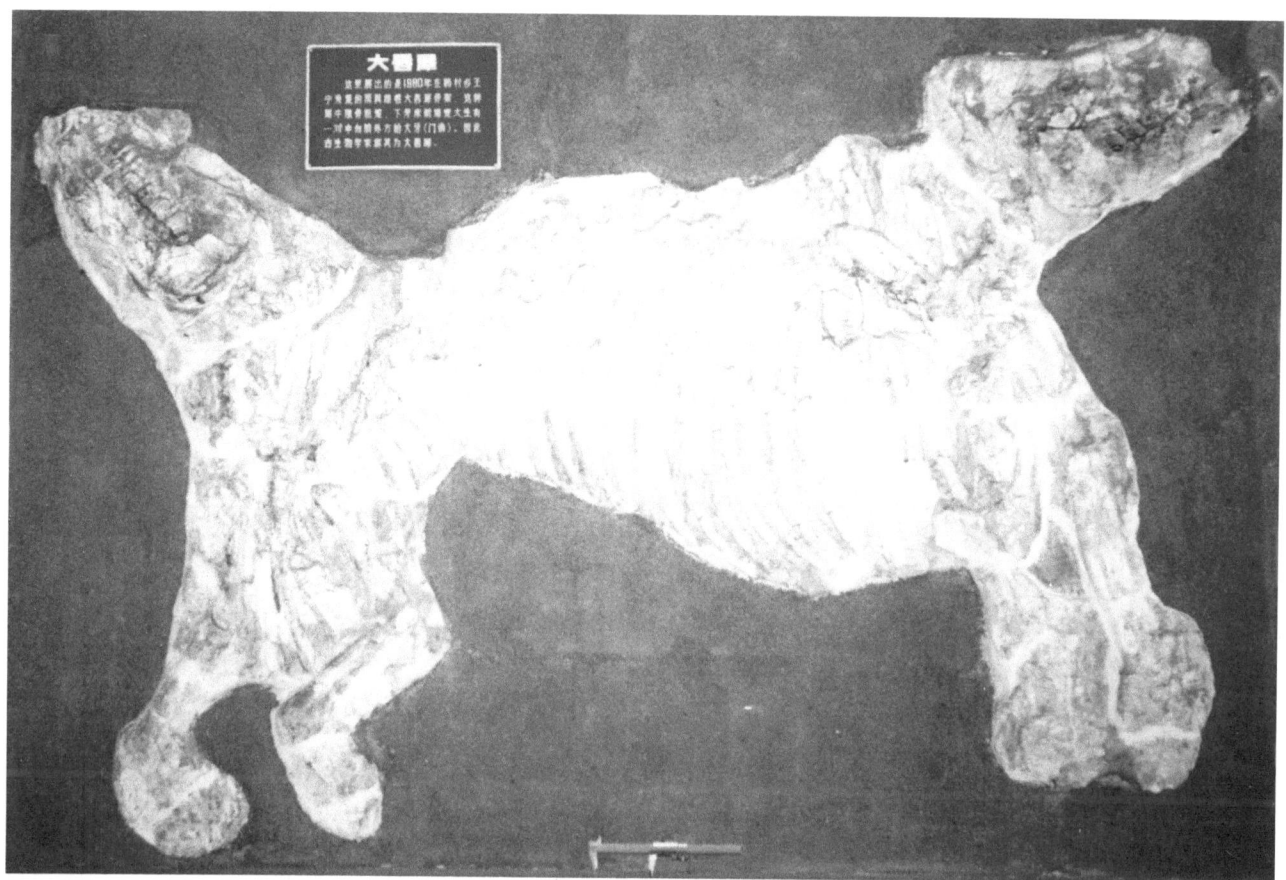

Fig. 2.14 *"Chilotherium"* skeletons found near Wangning village by Cao's party in 1980. The slab is currently mounted in the Yushe County Museum

during his stay in Germany (1982–1984) of the hyaenid specimens kept in IVPP. Heintz Tobien, Guan-Fang Chen and Yu-Qing Li undertook a revision of the Yushe Proboscidea (Tobien et al. 1986, 1988).

Paleontological study by the Institute of Geology and Paleontology, Nanjing. From 1979 to 1985 several groups of invertebrate paleontologists and paleobotanists intermittently collected fossils in the late Cenozoic deposits in middle and southern Shanxi. Their results were compiled into a special volume, and presented at the INQUA Congress XIII held in Beijing (Huang and Guo 1991). Authors of this volume included scientists of the Institute of Geology and Paleontology, Nanjing, Academia Sinica (Bao-Yu Huang, Bao-Ren Huang, Zhen Wang, Ling-Yu Tang), North China Bureau of Oil Geology, Ministry of Geology and Mineral Resources of China ([now Ministry of Land and Resources of China] Shu-Yuan Guo, Jing-Zhe Wang, Ze-Run Zhang), Shanghai Natural History Museum (Hui-Ji Wang), and Institute of Geography and Limnology, Nanjing, Academia Sinica (Rui-Jin Wu). Although their study areas covered a vast territory of middle and southeastern parts of Shanxi Province, special emphasis was laid

on description of the invertebrate and plant fossils they discovered in the Yushe Group. The most remarkable alteration in stratigraphic division they suggested was discarding as distinct the Louzeyu Formation (they included it in the Zhangcun Formation), and the creation of a new unit, the Gengxiu Formation, based on a suite of siltstones and clays with rich ostracod fossils. About 30 m thick, the Gengxiu Formation unconformably overlies the Yushe Group but underlies the Lishi loess in a section 500 m north of the village Gengxiu (formerly Houmu) in the Ouniwa Subbasin. They also believed that this suite of siltstones and clays produced the derived zokor (originally *Siphneus* cf. *arvicolinus*, revised as *Allosiphneus arvicolinus*) fossils reported by Teilhard de Chardin and Young (1933). As a result, they thought that the age of the Gengxiu Formation of the Ouniwa Subbasin could prove to be older than that of the upper member of the Zhangcun Formation ("Louzeyu Formation") in the Zhangcun Subbasin, but it would still be a basal Pleistocene unit, correlatable with Licent and Trassaert's Zone 3.

More recently, Shi (1994) resurrected the Louzeyu Formation and redefined it based on facies analysis and

Fig. 2.15 Photo taken in the yard of the Yushe County Guest House in October 1980. *Front row* H. Tobien (*center*), Zhou Ming-Zhen (to his *right*, *third* from the *left*), Zhan-Xiang Qiu (*second* from *left*); K. Heissig (*third* from *right*), *Back row from left to right* Zhi-Hui Guo (TNHM), Wei-Long Huang (TNHM), De-Fa Yan (IVPP), Zhe-Ying Chen (Shanxi Institute of Archeology), Hang Jia (IVPP), Guan-Fang Chen (IVPP), Yu-Qing Li (TNHM)

mineralogical differences between the Zhangcun and Louzeyu Formations. Based on his paleomagnetic data, Shi Ning roughly compared the age of the Louzeyu Formation with that of Qiu and colleagues' Haiyan Formation, both early Pleistocene deposits (reversely magnetized, early Matuyama Chron). Following upon the work of Shi (1994), a Nanjing Institute team reanalyzed the palynological record of the Zhangcun Formation and found significant environmental change just following the Gauss-Matuyama boundary (Liu et al. 2002).

Interest in the Yushe Basin by the Global Community. By the mid-1980s the Yushe Basin became well known to vertebrate paleontologists for its wealth of fossils from numerous horizons, which would lend to creating a real local biostratigraphy. Among the foreign paleontologists who visited Yushe, Prof. H. Tobien and Dr. K. Heissig came from Germany in October 1980 (Fig. 2.15). In 1982, Dr. R. H. Tedford (May), Dr. L. Ginsburg of Muséum National d'Histoire Naturelle, Paris, (September), and

Dr. A. Forstén from the University of Helsinki, Finland (October) all saw basin deposits. Later (1987) Prof. J. Desmond Clark of the University of California, Los Angeles (May), and Drs. S. Mahmood Raza and I. U. Cheema representing the Geological Survey of Pakistan and the Pakistan Museum of Natural History (September) visited the area.

2.3 Field-Based Research of the Sino-American Yushe Project (SAYP, 1987–1998)

The idea to initiate a Sino-American joint project to combine a thorough restudy of the late Cenozoic fossil mammals held in China (Licent Collection) and the USA (Frick Collection), coupled with field work and techniques used in present-day bio- and chrono-stratigraphy, was first put forward by Richard H. Tedford as early as 1981 (letter of January 23, 1981, to Prof. Chow Minchen (Ming-Zhen Zhou), then director of IVPP). While talking with Dr. Chuan-Kui Li, a top specialist on Chinese Neogene micromammals who was visiting the AMNH in January 1981, Tedford mentioned the fossils from Shanxi localities in the Frick Collection and expressed his hope to develop a joint project with Chinese colleagues. Tedford roughly delineated the possible future field work, emphasizing magnetostratigraphy and the related mandatory mapping and section measuring in the most important basins in Shanxi Province. While attending the "*Hipparion* Workshop" held at the AMNH from 2 through 10 of November 1981 (see Eisenmann et al. 1988), Tedford and Zhan-Xiang Qiu had many good conversations about the details to develop a joint project. In May 1982, Tedford was invited to China so that he could get acquainted with the IVPP Neogene Group and explore the details of joint work in the field and laboratory. Accompanied by Qiu, Tedford spent a couple of days visiting southeastern Shanxi (Fig. 2.16). Unfortunately, the scheduled visit to the Shouyang Basin, the origin of an important part of the Shanxi fossils in the Frick Collection, failed because of unexpected difficulties in getting special permission for foreigners to go there at that time. However, Tedford enjoyed the field experiences in Yushe and Wuxiang counties, and his imagination for the potential for biostratigraphic studies was captivated.

Qiu was granted a Humboldt scholarship for one and a half years' stay at Gutenberg University in Mainz (June 1982, to the end of 1983), and Tedford had an Australian project during 1983, Tedford and Qiu agreed to postpone organization of the project until 1984.

Qiu finished his Humboldt scholarship in June 1984. En route back to Beijing, Qiu was invited to visit the AMNH in early July. During this short stay in New York, Tedford and Qiu spent much time on details of the preparation of the joint project proposal. The latter half of 1984 was a busy time for Tedford to draw up the first draft of the proposal, which was ready in August of that year. Gradually we realized that the joint project should be confined to the Yushe-Wuxiang Basin. Meanwhile the final list of the participants of the project had been fixed. The formal grant proposal was completed in August 1985, and was received on October 10 by the NSF (USA). Unfortunately, the project was rejected by NSF reviewers who thought that the project had "too many goals," unrelated to the "important questions of ape and human origins," and demanded too large a budget. However, we were encouraged to submit the proposal again, with a smaller budget and narrower project. At last, on July 7, 1987, the SAYP was launched by a formal notice of the acceptance of the grant proposal EAR 8709221 under the title "Neogene Rocks and Faunas,

Yushe Basin, Shanxi, PRC" (July 1987, through June 1989) was received. The project was followed by another grant under the title "Evolution and Faunal Turnover in the Neogene of Northeastern Asia" (July 1991, through June 1993).

2.3.1 Field Campaigns

Under the auspices of "Neogene Rocks and Faunas, Yushe Basin, Shanxi, PRC" intensive field work began, mainly in the Yuncu subbasin in the first 2 years. Subsequent work by SAYP documented the Tancun subbasin, added observations throughout Yushe Basin, and complemented these studies with reconnaissance in Wuxiang Basin.

1987 (September 4–October 3)

Participants: R. H. Tedford, L. J. Flynn, N. D. Opdyke, W. R. Downs, Zhan-Xiang Qiu, Wen-Yu Wu, De-Fa Yan, Jie Ye, Xiao-Feng Chen (postgraduate student), Yi-Zheng Li (postgraduate student), Gen-Zhu Zhu (assistant curator of the IVPP Museum), Tai-Ming Wang (Yushe County Museum; Fig. 2.17).

Geology: Study was concentrated on two basal formations, the Mahui and the lower part of the Gaozhuang (Fig. 2.18). About 20 km^2 of Yuncu subbasin were mapped in reconnaissance fashion; five long sections plus several shorter ones were measured: CAN (Nanmahui), CAB (Beimahui), CAL (Lintou), CAH (Shagou), CAJ (Jingjiagou) and CAG (Gaozhuang).

Paleontology: Prospected Lintou, Gaozhuang, and Nanzhuanggou areas of Yuncu subbasin, with 77 localities catalogued (YS1 through YS77). Large mammalian fossils: With the help of the local people most of Licent's and IVPP 1955–1956 localities in the Yuncu subbasin and their stratigraphic position were settled. Micromammals: Surface finds of skull and jaw material of small mammals at different levels of the Mahui and Gaozhuang Formations indicated that the possibility in finding richer micromammal fossils in the future was very high. Ten micromammal sites were sampled by screen washing and the concentrate was partly sorted in the field.

Magnetostratigraphy: Paleomagnetic sampling from fine-grained sediments of the Mahui and Gaozhuang Formations were systematically conducted by Neil Opdyke and his assistants at approximately 3.5–5 m intervals.

1988 (September 4–October 9)

Participants: The Americans were the same as in 1987; from the Chinese side: Zhan-Xiang Qiu, Zhu-Ding Qiu, De-Fa Yan, Wen-Yu Wu, Guan-Fang Chen, Yu-Qing Li (TNHM), plus from Mainz, Germany, for 1 week, Norbert Schmidt-Kittler (Figs. 2.19, 2.20).

Fig. 2.16 R. H. Tedford (*right*) and Zhan-Xiang Qiu (*front left*) observing a *Stegodon* skull at Yushe County Museum in May 1982

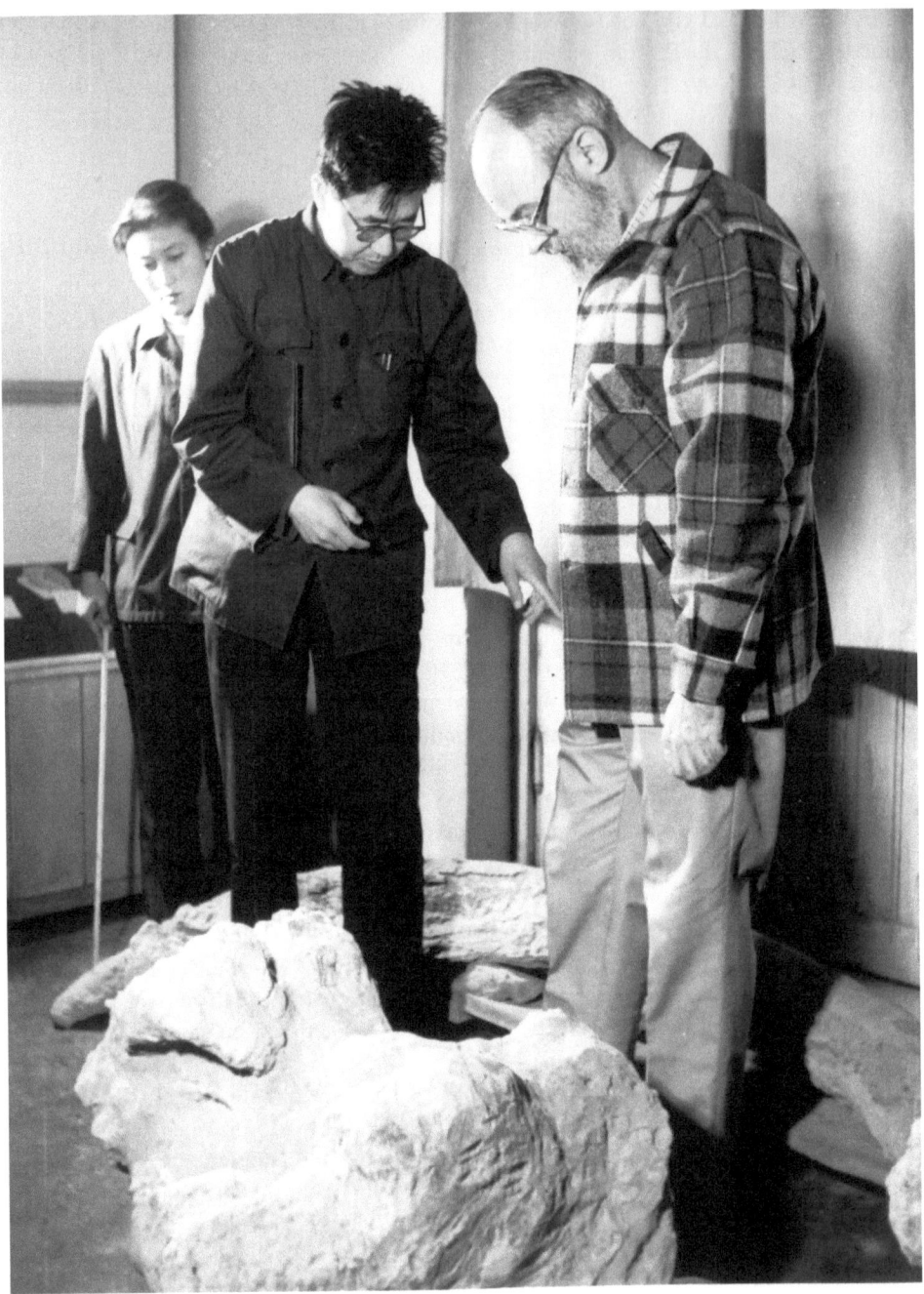

Geology: The upper part of the Gaozhuang Formation, the Mazegou and Haiyan Formations were studied. The entire stratigraphic 800 m column was assembled using altogether 25 sections. The mapping of the entire Yuncu subbasin (about 300 km^2) was completed at 1:50,000 scale.

Paleontology: Prospected areas included Taoyang, Baihai, Nanzhuanggou, Zhaozhuang, Mazegou, Malan, Liujiagou, Songyangou, and other parts of Yuncu subbasin. Most of Licent's richly fossiliferous sites of Pliocene and Pleistocene ages were located. Screen washing about 30 sites produced concentrate for sorting, which yielded new small mammal samples. Altogether about 60 new vertebrate localities were catalogued (YS78 through YS137).

Magnetostratigraphy: Systematic sampling was conducted in Mazegou and Haiyan Formations.

1991 (September 13–30)

Participants: R. H. Tedford, L. J. Flynn, W. Downs, Jie Ye, Zhu-Ding Qiu, Wei Dong, Yu-Qing Li, Guan-Fang Chen, Xiaoming Wang and Tai-Ming Wang (Fig. 2.21).

Field work: Systematic prospecting and fossil collection in the Tancun subbasin, with relocation of the type locality

Fig. 2.17 Part of the team welcomed by local people in front of the Yushe County Guest House in September 1987. *Front row* Xiao-Feng Chen (IVPP, *third left*), Wen-Liang Jia (Yushe County Museum, *fifth left*), Zhan-Xiang Qiu (IVPP, *fourth right*), W. R. Downs (*third right*), Wen-Yu Wu (IVPP, *second right*). *Second row* De-Fa Yan (IVPP, *first left*), Gen-Zhu Zhu (IVPP, *third left*), Jie Ye (IVPP, *fourth left*). *Third row, left* Tai-Ming Wang (Yushe County Museum)

Fig. 2.18 Trio of team geologists (*left* to *right* Zhan-Xiang Qiu, R. H. Tedford, and Jie Ye) taken in front of the Yushe County Guest House in September 1987

Fig. 2.19 Inside the Yushe County Museum. Wen-Liang Jia, Zhan-Xiang Qiu and R. H. Tedford examine specimens being prepared for display (*photo* L. Flynn, September 1987)

for *Neocricetodon grangeri*, and screening at several localities below and above the Mahui Formation-Gaozhuang Formation contact. Localities YS138 through YS171 were found and catalogued.

Magnetostratigraphy: Four sections were measured and sampled in the Tancun subbasin, CAW+CAWL (Sijiawa) and CAU+CAUL (Jiayucun). In Yuncu subbasin, magnetic samples were added to constrain the Early Pleistocene CAQ section at Qingyangping.

1994 (October 5–11)

Participants: R. H. Tedford, Zhan-Xiang Qiu, Xiaoming Wang, Jie Ye, Wen-Yu Wu, and Tai-Ming Wang.

Field work: Cursory survey of the Yushe Group in the Zhangcun subbasin was carried out. The Renjianao, Wangning and Zhangcungou areas were prospected. Prospecting and tracing the boundary between the Triassic rocks and the Yushe Group in Jiayucun, Dengyucun, Taiqu areas of the Tancun subbasin, and the Liutan area of the northern Tancun subbasin were also conducted.

1997 (August 9–15)

Participants: Zhan-Xiang Qiu, Wei-Long Huang (TNHM) and Tai-Ming Wang.

Field work: The main purpose of the year's trip was to visit Licent's famous Loc. 2, Zhangcungou (Wuxiang County) of the Zhangcun subbasin, which Licent used as the main base for his geologic survey and excavation, and as center of his fossil purchasing activities, for more than 3 weeks, from July 3 to 26 in 1934. With the help of the local people, we located the exact place where Licent made his excavation and found the section he used to illustrate his

subdivision of the deposits of this area. Later, Haobei, Dongfangshan and Danangou of the Tancun subbasin, Dazhai and Nanhedi of the Ouniwa subbasin, and Yimen of the Wuxiang Basin were also visited.

While surveying Zhangcungou village, we made inquiries about Licent's collecting activities in this area. Three old villagers (Fig. 2.22) provided us with the following information. There were two foreign fathers (a yellowish and a blackish), apparently Licent and Trassaert, accompanied by two bodyguards (Wang and Tian) with seven mules. They hired a "dragon bone" dealer, Hao Lin-Zhong (then middle-aged) from Shibi village of the Wuxiang County, as their agent in charge of collecting and purchasing fossils. The foreigners made some excavations mainly in a gully near Xiongshujia village (no longer extant).

The cliff in the gully near Xiongshugou is apparently the section presented by Licent and Trassaert in 1935 and later by Huang and Guo in 1991 (vide supra, Fig. 2.12, field image in Fig. 2.23). We revisited this gully and found a *Paramachairodus* mandible in the upper part of that section, indicating an age comparable with that of the Mazegou Formation in the Yuncu subbasin.

1998 (October 8–13)

Participants: R. H. Tedford, Zhan-Xiang Qiu, Hong Zhao (assistant curator of the IVPP Museum) and Tai-Ming Wang.

Field work: The main purpose of the field work was to investigate the deposits of the Ouniwa subbasin, especially the thick basal conglomerate layers near Nanhedi.

Supplementary surveys and reconnaissance were conducted by different groups for all subbasins in the years 1991,

Fig. 2.20 Attending the dedication ceremony of the Yushe County Museum in September 1988. Sitting behind the desk, from *right* to *left* W. Downs, L. Flynn, N. Schmidt-Kittler, N. Opdyke, R. Tedford, Z.-X. Qiu, Y. Zhang (Vice-director of Bureau of museums and archeology of Shanxi Province), and other officials

1994 and 1997–1998, with the aim to get a better understanding of the geological history of the Yushe Basin as a whole.

2.3.2 Publications

Concomitantly with the field work, short papers concerning field and laboratory results appeared during the interval 1989–1996. A dozen titles are listed here in chronological order:

1. Tedford, R.H., Flynn, L.J., Qiu, Z.-X., 1989. Neogene faunal succession, Yushe Basin, Shanxi Province, PRC. Journal of Vertebrate Paleontology 9 (supplement to No. 3), 41A.
2. Flynn, L.J., Tedford, R.H., Qiu, Z.-X., 1990. The Yushe chronofauna: faunal stability in the Pliocene of North China. Journal of Vertebrate Paleontology 10 (supplement to No. 3), 23A.
3. Qiu, Z.-X., Tedford, R.H., 1990. A Pliocene species of *Vulpes* from Yushe, Shanxi. Vertebrata PalAsiatica 28 (4), 245–258.
4. Tedford, R.H., Flynn, L.J., Qiu, Z.-X., Opdyke, N.D., Downs, W.R., 1991. Yushe Basin, China: paleomagnetically calibrated mammalian biostratigraphic standard for the Late Neogene of eastern Asia. Journal of Vertebrate Paleontology 11 (4), 519–526.
5. Tedford, R.H., Qiu, Z.-X. (1991). Pliocene *Nyctereutes* (Carnivora: Canidae) from Yushe, Shanxi, with comments on Chinese fossil raccoon-dogs. Vertebrata PalAsiatica 29 (3), 176–189.
6. Flynn, L.J., Tedford, R.H., Qiu, Z.-X., 1991. Enrichment and stability in the Pliocene mammalian fauna of North China. Paleobiology, 17 (3): 246–265.

Fig. 2.21 Part of the 1991 Sino-American Yushe team in front of the Yushe Guest-House. *Seated from right* Tai-Ming Wang, Will Downs, Dick Tedford, Larry Flynn; *standing from right* Guan-Fang Chen, Yi-Zheng Li, Jie Ye, Zhu-Ding Qiu, Wei Dong

7. Wu W.-Y., Flynn, L.J., 1992. New murid rodents from the Late Cenozoic of Yushe Basin, Shanxi. Vertebrata PalAsiatica 30(1), 17–38, 2 pl.

8. Flynn, L.J., 1993. A new bamboo rat from the late Miocene of Yushe Basin. Vertebrata PalAsiatica 31, 95–101.

9. Flynn, L.J., Wu W.-Y., 1994. Two new shrews from the Pliocene of Yushe Basin, Shanxi Province, China. Vertebrata PalAsiatica 32 (2), 73–86.

10. Flynn, L.J., Qiu, Z.-X., Opdyke, N.D., Tedford, R.H., 1995. Ages of key fossil assemblages in the Late Neogene terrestrial record of northern China. Geochronology Time Scales and Global Stratigraphic Correlations. SEPM (Society for Sedimentary Geology) Special Publication 54, 365–373.

11. Tedford, R.H., 1995. Neogene mammalian biostratigraphy in China: past, present, and future. Vertebrata PalAsiatica 33 (4), 272–289.

12. Tedford, R.H., Qiu, Z.-X., 1996. A new canid genus from the Pliocene of Yushe, Shanxi Province. Vertebrata PalAsiatica 34 (1), 27–40.

2.4 Problems Related to the Historical Fossil Collections

The Wade-Giles romanization of the Chinese geographic names used in the past produced a considerable degree of misunderstanding that any reader who skims through the present series will acutely feel. Fortunately, the Pinyin Romanization System was officially decreed by the Chinese government in 1979. Since then the new system has widely been accepted in China and abroad. In order to avoid further misunderstanding, the Pinyin Romanization System is applied throughout the present volume. The old names

Fig. 2.22 Some of the Zhangcungou villagers consulted by Qiu in August 1997. *Front, from left to right* De-Jun Hao (1911?–), (Z.-X. Qiu), Ming-Wen Hao (1926–), Ming-Zhong Hao (1932?–)

using the Wade-Giles system are avoided as far as possible, and when necessary to use, they are written in italics.

The fossil collections studied in the course of our research comprise altogether seven parts, the first four of which were obtained during the 1920s–1930s. These are: (1) the unstudied fossils of the Lagrelius Collection obtained by Andersson's assistants during the 1920s from the Yushe-Wuxiang area, which are now kept in IVPP; (2) the major part of the Licent Collection held in Tianjin Natural History Museum (catalogued as THP and TNP); (3) a small portion of the Licent Collection now housed in IVPP (catalogued as RV or V); (4) the Childs Frick Collection kept in the American Museum of Natural History (catalogued as F:AM); (5) the specimens collected by the Laboratory of Vertebrate Paleontology in the 1950s, now housed in the IVPP; (6) a small number of specimens in the Yushe County Museum; and (7) the few specimens obtained by the Neogene Division of IVPP since 1979–1980 (catalogued as QY), plus the larger collection built during the

execution of the Sino-American Yushe Project from 1987 to 1998 (catalogued as YS), now housed in the IVPP.

The Licent Collection comprises the bulk of the specimens studied in the present series. The total number of these specimens is about 2,300. At first, these fossils were not regularly numbered or catalogued in situ. On some of these fossils, serial numbers and dates were written in ink. For instance, "5, 4/VII, 1934" means specimen No. 5, collected on July 4, 1934. These specimens with serial numbers and dates serve as a good check on the localities, since they were entered in Licent's field notes. All the fossils collected by Licent's parties in 1934–1935 were systematically catalogued later in the *Musée Hoang-ho Pai-ho de Tientsin* in 1935 or 1936. The numbering system started from 10,000 without prefix. They were apparently casually numbered, without systematic or locality consideration. So, neighboring numbers can be given to fossils belonging to different taxa and different localities. The person in charge of the cataloguing (probably Father Haser, who was said to know

Fig. 2.23 Landscape of Licent's Loc. 2 (*Changts'unkou*), northwest view, photographed by Qiu in August 1997. X: site of Licent's 1934 excavation. One of the two gullies southeast of the excavation site would be where Licent and Trassaert measured their section in 1934

both French and Chinese) apparently did not participate in the field work and did not know these localities in the field, because there are various incorrect assignments and trans-literations of the Chinese characters for the localities. The following are the most obvious mistakes:

(1) Loc. 15, *Ichuangts'un* is an incorrect transliteration from French to Chinese of the Loc. 4, *Chaochuangts'un* (Zhaozhuang village), since the simplified and cursive-hand Chinese character *Chao* (Zhao in Pinyin Roman-ization) looks very much like a Chinese character Yi.

(2) Loc. 14, *Hsingyangts'un* and Loc. 18, *Laohsiangts'un* are both incorrect transliterations of the Loc. 44, *Tao-yangts'un* (Taoyang village).

(3) Loc. 16, *Kaochaungts'un* is the same as Loc. 15, *Kaochuang* (Gaozhuang).

(4) Loc. 49, *Chaments'un* is Loc. 63, *Yiments'un* (Yimen village, see Appendix IV).

(5) Loc. 57, *Hsuhochangts'un* may be the wrong translit-erations of the *Ni Ho Chang* (Nihezhang).

(6) Loc. 60, *Litats'un* is Loc. 20, *Liyuts'un* (Liyu village).

(7) Loc. 61, *Soanhots'un* may be Loc. 25, *Nihots'un* (Zhongnihe village).

While describing the first skull of *Plesiohipparion hou-fenense* of the Licent Collection (Qiu et al. 1980) Qiu and Wei-Long Huang, then in charge of the curation of the vertebrate fossils in the Tianjin Natural History Museum, felt the inconvenience of using the numbering system without an institutional prefix. They proposed to add the prefix THP (T for Tianjin, H for *Hoang-ho Pai-ho*, P for Paleontology) to the given numbers pertaining to the Licent Collection, and TNP (N for Natural History) for the speci-mens other than the Licent Collection. A small number of specimens of the Licent Collection lost their original numbers for unknown reasons. For them new TNP numbers were given. In many cases the provenances of these speci-mens are unknown.

The catalogue numbers for the studied specimens now housed in the IVPP contain serial numbers prefixed with V. This system has been in use since 1937, when the Cenozoic Research Laboratory moved to Sichuan Province after the Anti-Japanese War broke out. Prior to 1937, no unified catalogue system existed. For instance, the specimens briefly described by C. C. Young in 1935 were not sys-tematically catalogued. Sometimes, locality numbers can be found written on the specimens. A few Yushe specimens found from 1937 to 1949 were catalogued with prefix V in the old collection at the IVPP (for example, V 517–526, see Young and Liu 1948). During the wartime, a small part of the Licent Collection was transferred from Tianjin to Beijing, temporarily housed in the Institut de Géo-biologie. After founding of the People's Republic of China in 1949, these specimens became the property of the Laboratory of Vertebrate Paleontology, the predecessor of the Institute of Vertebrate Paleontology and Paleoanthropology, to which these transferred specimens now belong. Working for both the Cenozoic Research Laboratory and the *Musée Hoang-ho Pai-ho de Tientsin*, Teilhard de Chardin used the number system without prefix regardless of where the specimens were housed. The studied specimens now housed in the IVPP are given new numbers prefixed by RV (R for *Re-catalogued*) and the first two numbers denote the year, with their original THP numbers in brackets. For instance, RV 4503 (THP 10552) means the specimen was studied and published in 1945. It was from the Licent Collection, but now belongs to the IVPP. Unstudied specimens of the Licent Collection now kept in the IVPP are given a serial number with prefix V when they are studied and published, with their original THP numbers in brackets.

The specimens procured in 1955–1956 by the Laboratory of Vertebrate Paleontology, Academia Sinica, bear only field locality numbers (for instance, 5544, 5679). In 1979, prior to the Sino-American Yushe Project, while prospecting the Yushe area, Qiu and Huang collected and purchased a number of specimens. They bear the locality numbers prefixed with QY (see Appendix VII). All the specimens procured over the course of the SAYP since 1987 bear unique locality numbers with the prefix of YS. When studied, these specimens are given catalog numbers prefixed by V.

Since many of the specimens, often the best-preserved skulls and jaws, were purchased from local villagers, their provenance is either lacking, or in cases with locality information, serious verification is needed. The general strategies we adopted are as follows. The first is to use as far as possible the fossils proven to be found in situ. Sometimes, these specimens may be unimportant for systematic study, but sufficient to prove the existence of certain forms in time and space. The second is to use as far as possible the degree of congruence between the mode of preservation, color, hardness, completeness, etc. of the fossils and the lithology of the designated locality. The third is to use preferably localities remote from the dragon-bone purchase centers, often a large village or town (zhen pertains to township in Chinese, while "zhuang" and "cun" correspond to village). Places such as Zhangwagou ("gou" is gully in Chinese) or Shennan'ao ("ao" is depression in Chinese), are more likely real fossil localities.

Our goals were to reconstruct locality information as accurately as possible to tie all collections to the stratigraphy, be they the fruits of historical explorations in the area, or new findings in the field. This maximizes the value of Yushe Basin fossils for biostratigraphy, and hopefully for recognition of characteristic fossil assemblages. Ultimately our goal is to apply all observations to test the durations of paleofaunas, recognize faunal turnover, and refine Late Neogene biochronology.

References

Andersson, J. G. (1919). Dragon-hunting in China. *Far Eastern Review, 1919*(November), 1–11.

Andersson, J. G. (1922). Current palaeontological research in China. *Bulletin of the American Museum of Natural History, 46*(3), 727–737.

Andersson, J. G. (1923). Essays on the Cenozoic of northern China. *Memoir of Geological Survey of China, A3,* 1–152.

Andersson, J. G. (1934). *Children of the Yellow Earth*. London: Kegan Paul, Trench, Trubner & Co.

Andrews, R. C. (1932). *Natural history of Central Asia, vol. 1. The new conquest of Central Asia. A narrative of the explorations of the Central Asiatic Expeditions in Mongolia and China, 1921–1930.* New York: American Museum of Natural History.

Bohlin, A. B. (1926). Die Familie Giraffidae mit besonderer Berücksichtigung der fossiler Formen aus China. *Palaeontologia Sinica C, 4*(1), 1–208, 12 plates.

Bohlin, A. B. (1935). Cavicornier der *Hipparion*-Fauna Nord-Chinas. *Palaeontologia Sinica C, 9*(4), 1–66.

Cao, J.-X. (1980). A study of Cenozoic strata and depositional environment of the Taigu-Yushe-Wuxiang area, Shanxi. *Quaternaria Sinica, 4*(1), 77–82 (in Chinese).

Cao, J.-X., & Cui, H.-T. (1989). Research of Pliocene flora and paleoenvironment of Yushe Basin on Shanxi plateau, China. *Scientia Geologica Sinica, 4,* 367–375.

Cao, J.-X., & Wu, R.-J. (1985). The characteristics of sediments and landform evolution of the late Cenozoic down-warped basin in Yushe and Wuxiang district, Shansi. *Quaternaria Sinica, 6*(2), 48–54 (in Chinese).

Cui, Z.-K., & Wu, T.-S. (2002). Geologic map of Shanxi. In L.-F. Ma (Ed.), *Regional geologic maps of China* (pp. 133–140). Beijing: Geological Publishing House.

Eisenmann, V., Alberdi, M. T., De Giuli, C., & Staesche, U. (1988). Volume I: Methodology. In M. O. Woodburne & P. Y. Sondaar (Eds.), *Studying fossil horses* (pp. 1–71). Leiden: E. J. Brill.

Geological Bureau of Shanxi Province. (1976). *Geologic Fenyang-Pingyao Quadrangle*. Beijing: Geological Publishing House.

Hopwood, A. T. (1935). Fossil Proboscidea from China. *Palaeontologia Sinica C, 9*(3), 1–108.

Hu, C.-K. (1962). A new species of *Metacervulus* of Yushe, Shansi, with notes on Pliocene muntjaks of China. *Vertebrata PalAsiatica, 6*(3), 251–261 (in Chinese with English summary).

Huang, B.-Y., & Guo, S.-Y. (1991). Stratigraphy. In B.-Y. Huang & S.-Y. Guo (Eds.), *Late Cenozoic stratigraphy and paleontology from Central-Southern region of Shanxi* (pp. 1–70). Beijing: Science Press.

Jacobs, L. L., & Li, C.-K. (1982). A new genus (*Chardinomys*) of murid rodent (Mammalia, Rodentia) from the Neogene of China, and comments on its biogeography. *Géobios, 15*(2), 255–259.

Jia, L.-P., & Zhen, S.-N. (1978). Dragon and dragon bone. In Editorial board of the Journal "Fossils" (Eds.), *Fossil World* (pp. 132–137). Beijing: Science Press (in Chinese).

Li, S.-Z. (1596). *Compendium of Materia Medica* (1982 edition). Beijing: People's Hygiene Publishing House.

Licent, E., & Trassaert, M. (1935). The Pliocene lacustrine series in central Shansi. *Bulletin of the Geological Society of China, 14*(2), 211–219.

Liu, G., Leopold, E. B., Liu, Y., Wang, W., Yu, Z., & Tong, G. (2002). Palynological record of Pliocene climate events in North China. *Review of Paleobotany and Palynology, 119*, 335–340.

Liu H.-T. (=Liu X.-T.), & Su T.-T. (=Su D.-Z.). (1962). Pliocene fishes from Yüshe Basin, Shansi. *Vertebrata PalAsiatica 6*(1), 1–26 (in Chinese with English summary).

Liu, X., Guo, P., & Liu, X. (2009). *Shan Hai Jing*. Shenyang: Wanjuan Publishing Company (in Chinese).

Mateer, N. J., & Lucas, S. G. (1985). Swedish vertebrate palaeontology in China: A history of the Lagrelius Collection. *Bulletin of the Geological Institutions of the University of Uppsala New Series, 11*, 1–24.

Needham, J. (1959). *Mathematics and the sciences of the heavens and the earth (Science and Civilisation in China 3)*. Cambridge: Cambridge University Press.

Pearson, H. S. (1928). Chinese fossil Suidae. *Palaeontologia Sinica, C5*(5), 1–75.

Pei, W.-Z., Zhou, M.-Z., & Zheng, J.-J. (1963). *Cenozoic Erathem of China*. Beijing: Science Press (in Chinese).

Qiu, Z.-X. (1987). Die Hyaeniden aus dem Ruscinium und Villafranchium Chinas. *Münchener Geowissenschaftliche Abhandlungen, A9*, 1–110.

Qiu, Z.-X., Huang, W.-L., & Guo, Z.-H. (1980). Notes on the first discovery of the skull of *Hipparion houfenense*. *Vertebrata PalAsiatica, 18*(2), 131–137 (in Chinese with English summary).

Qiu, Z.-X., Huang, W.-L., & Guo, Z.-H. (1987). The Chinese Hipparionine Fossils. *Palaeontologia Sinica, New Series, C25*, 1–243 (in Chinese with English summary).

Ringström, T. (1927). Über Quartäre und Jungtertiäre Rhinocerotiden aus China und der Mongolei. *Palaeontologia Sinica C, 4*(3), 1–23.

Schmitz-Moormann, K. (Ed.). (1971). *Pierre Teilhard de Chardin, L'Oeuvre Scientifique* (10 Vols., 4634 p). Freiburg: Walter-Verlag, Olten und Freiburg im Breisgau.

Sefve, I. (1927). Die Hipparionen Nord-Chinas. *Palaeontologia Sinica C, 4*(2), 1–91.

Shi, N. (1994). The Late Cenozoic stratigraphy, chronology, palynology and environmental development in the Yushe Basin, North China. *Striae, 36*, 1–90.

Teilhard de Chardin, P. (1942). New rodents of the Pliocene and Lower Pleistocene of North China. *Publications de l'Institut de Géobiologie, Pékin, 9*, 1–101.

Teilhard de Chardin, P., & Leroy, P. (1945a). Les Félidés de Chine. *Publications de l'Institut de Géobiologie, Pékin, 11*, 1–58.

Teilhard de Chardin, P., & Leroy, P. (1945b). Les Mustélidés de Chine. *Publications de l'Institut de Géobiologie, Pékin, 12*, 1–56.

Teilhard de Chardin, P., & Trassaert, M. (1937a). The Proboscideans of South-Eastern Shansi. *Palaeontologia Sinica C, 13*(1), 1–85.

Teilhard de Chardin, P., & Trassaert, M. (1937b). Pliocene Camelidae, Giraffidae and Cervidae of South-Eastern Shansi. *Palaeontologia Sinica New Series C, 1*, 1–69.

Teilhard de Chardin, P., & Trassaert, M. (1938). Cavicornia of South-Eastern Shansi. *Palaeontologia Sinica New Series C, 6*, 1–107.

Teilhard de Chardin, P., & Young, C. C. (1933). The late Cenozoic Formations of S. E. Shansi. *Bulletin of Geological Society of China, 12*, 207–248.

Tobien, H., Chen, G.-F., & Li, Y.-Q. (1986). Mastodonts (Proboscidea, Mammalia) from the Late Neogene and Early Pleistocene of the People's Republic of China. Part 1. *Mainzer Geowissenschaftliche Mitteilungen, 15*, 119–181.

Tobien, H., Chen, G.-F., & Li, Y. Q. (1988). Mastodonts (Proboscidea, Mammalia) from the Late Neogene and Early Pleistocene of the People's Republic of China. Part 2. *Mainzer Geowissenschaftliche Mitteilungen, 17*, 95–220.

Yang, Z.-J. (2009). *Section of section* (C.C. Young's field notes of 1932–1936, in Chinese). Beijing: Science Press.

Young, C. C. (=Yang Z.-J.). (1935). Miscellaneous mammalian fossils from Shansi and Honan. *Palaeontologia Sinica C, 9*(2), 1–42.

Young, C. C., & Liu, P. T. (=Liu D.-S.) (1948). Notes on a mammalian collection probably from the Yushe series (Pliocene), Yushe, Shansi, China. *Contributions from the Geological Institute, National University of Peiping 8*, 273–291.

Zdansky, O. (1925). Fossile Hirsche Chinas. *Palaeontologia Sinica C, 2*(3), 1–94.

Zdansky, O. (1927). Weitere Bemerkungen über fossile Carnivoren aus China. *Palaeontologia Sinica C, 4*(4), 1–30.

Chapter 3
Cenozoic Geology of the Yushe Basin

Richard H. Tedford, Zhan-Xiang Qiu, and Jie Ye

Abstract The Yushe Group of Yuncu subbasin constitutes a long series of superposed units that span a large portion of the Late Cenozoic Era from Late Miocene, through Pliocene, to Early Pleistocene time. Hiatuses, some represented by angular unconformities, represent up to a half million years of missing time, but are shorter lower in the section. In Yuncu subbasin, the composite section is 800 m in length and contains four stratigraphic units distinguished by distinctive cycles of sedimentation. The basal Mahui Formation is conglomeratic toward the base, with coarse sands fining upward and with capping marls. It yields Late Miocene age vertebrate fossils. The Gaozhuang Formation contains abundant sands and finer units, with three fining-upward cycles that are recognized as members. The Mazegou Formation, with basal disconformity, contains fine sands and relatively more silts. Paleomagnetic study coinciding with our fieldwork shows that Gaozhuang and Mazegou fossil horizons span much of the Pliocene Epoch. These units dip gently, but are overlain by Early Pleistocene horizontal lake and stream sediments of the Haiyan Formation. Other subbasins contain sediments equivalent to parts of this sequence, and we recognize the Mahui-Gaozhuang contact to the east in Tancun subbasin. The Zhangcun subbasin exhibits a different, more lacustrine mode of deposition, with developed oil shales bearing abundant fish faunas. The Yushe Group is mantled by loess, locally with an older deep red loess likely equivalent in part to the Early Pleistocene Wucheng Loess. More widespread are the younger, thick Lishi and Malan loess blankets. This chapter opens and closes with syntheses of basin stratigraphy, set against the structural and climatic evolution of Yushe Basin. The underlying structure of the Triassic bedrock dominates the tectonics of Yushe Basin. Sediment aggradation along the trunk of the Zhuozhang paleo-river of Yushe Basin began after 7 Ma. Progressive aggradation continued, and the hiatuses in deposition and minor angular unconformities indicate episodes of gentle uplift. As the basin filled, the later Pleistocene loess signaled a new mode of sedimentation.

Keywords Geology • Yushe Group • Mahui Formation • Gaozhuang Formation • Mazegou Formation • Haiyan Formation • Loess • Structure

3.1 Overview

The late Cenozoic continental sediments that constitute the deposits of the Yushe Basin lie in the upper reaches of the drainage of the Zhuozhang River and its tributaries in southeastern Shanxi Province. These rocks form a NNE-SSW oriented band 75 km long and 25 km wide covering an area of 1875 km^2 in Yushe, Wuxiang, and Qin counties (Fig. 3.1). Late Cenozoic deposits of the Yushe Basin are contained in a deeply incised dendritic drainage system initiated in pre-late Miocene time and reexcavated during the Quaternary. This ancient drainage dissects the major axis of a synclinorium developed in Triassic continental deposits whose limbs extend eastward to the Taihang Shan and westward to the Taiyue Shan where they lie on Mesozoic and older rocks. These mountains formed the ultimate drainage divides for the Neogene Yushe Basin. Faulting and jointing of the Triassic rocks, parallel to the NE-SW trends of the fold axes, played an important role in localizing the development of the valleys of the ancient rivers. The initiation of late

R. H. Tedford
Formerly Division of Paleontology, American Museum of Natural History, Central Park West at 79 St., New York, NY 10024, USA

Z.-X. Qiu (✉) · J. Ye
Laboratory of Paleomammalogy, Institute of Vertebrate Paleontology and Paleoanthropology, Chinese Academy of Sciences, Xizhimenwai Ave., 142, Beijing 100044, People's Republic of China
e-mail: qiuzhanxiang@ivpp.ac.cn

J. Ye
e-mail: yejie@ivpp.ac.cn

R. H. Tedford, Z.-X. Qiu, L. J. Flynn (eds.), *Late Cenozoic Yushe Basin, Shanxi Province, China: Geology and Fossil Mammals. Volume I: History, Geology, and Magnetostratigraphy,* Vertebrate Paleobiology and Paleoanthropology, DOI: 10.1007/978-90-481-8714-0_3, © Springer Science+Business Media Dordrecht 2013

(a)

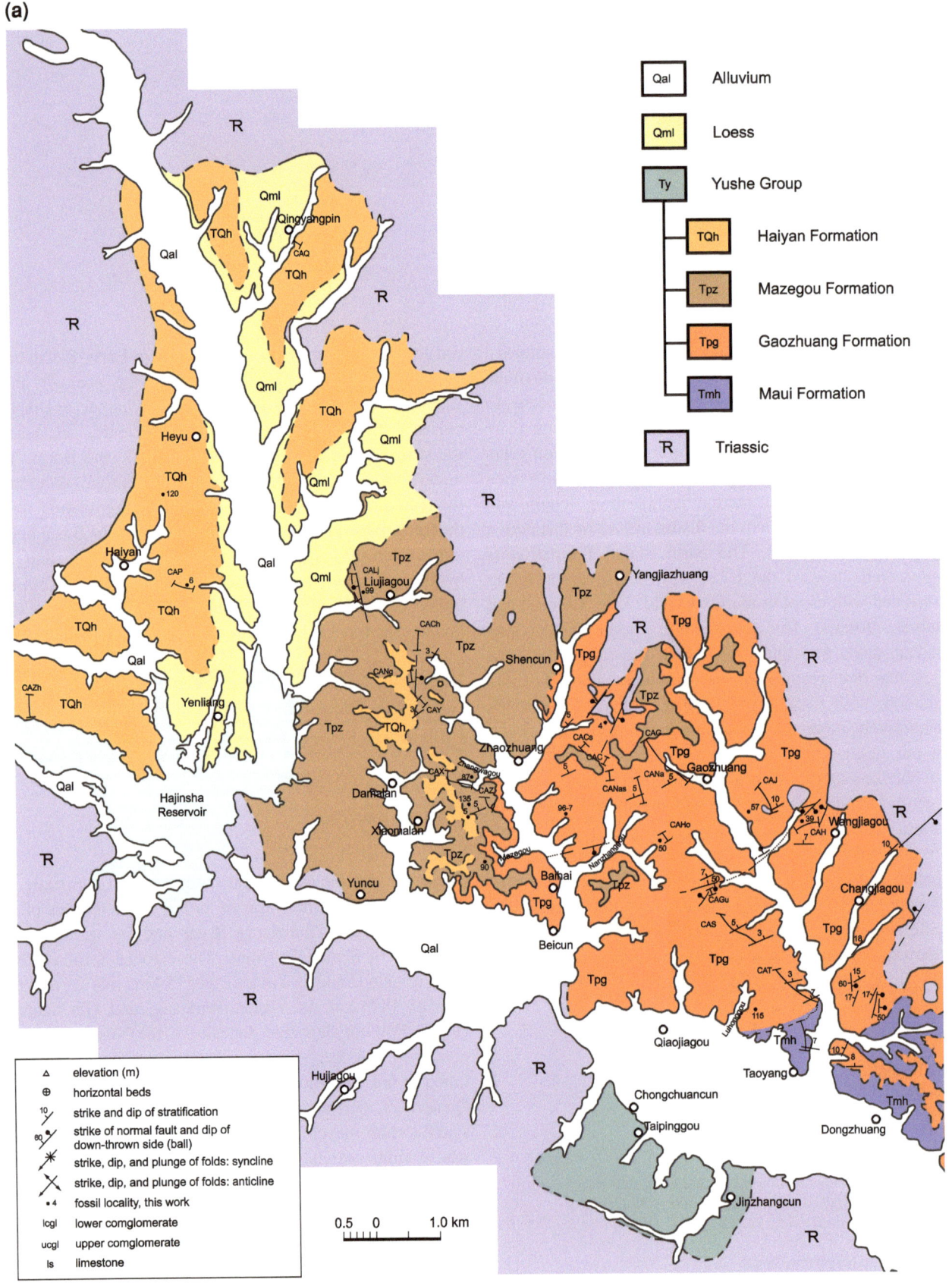

Fig. 3.1

(b)

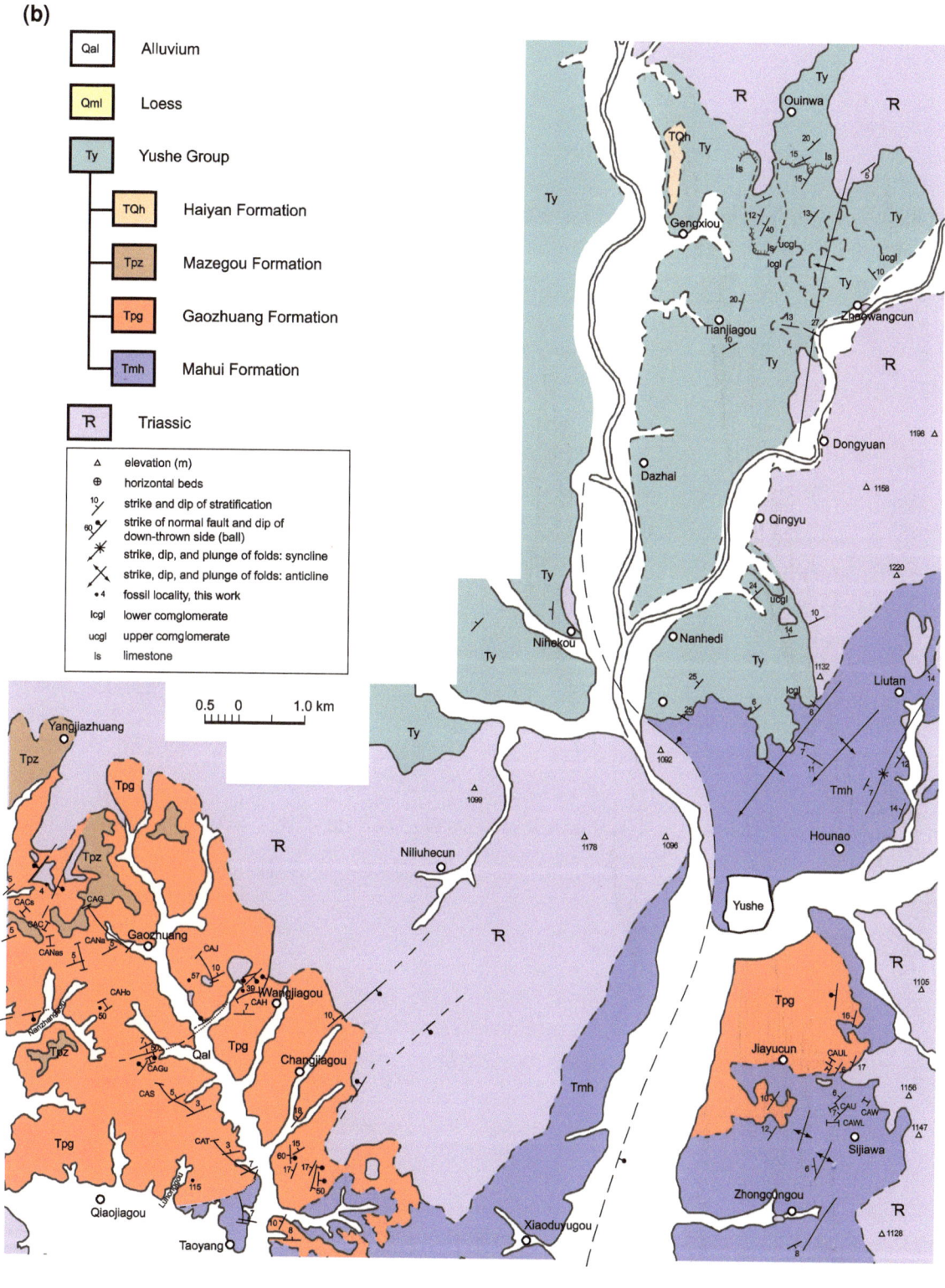

Fig. 3.1 (continued)

(c)

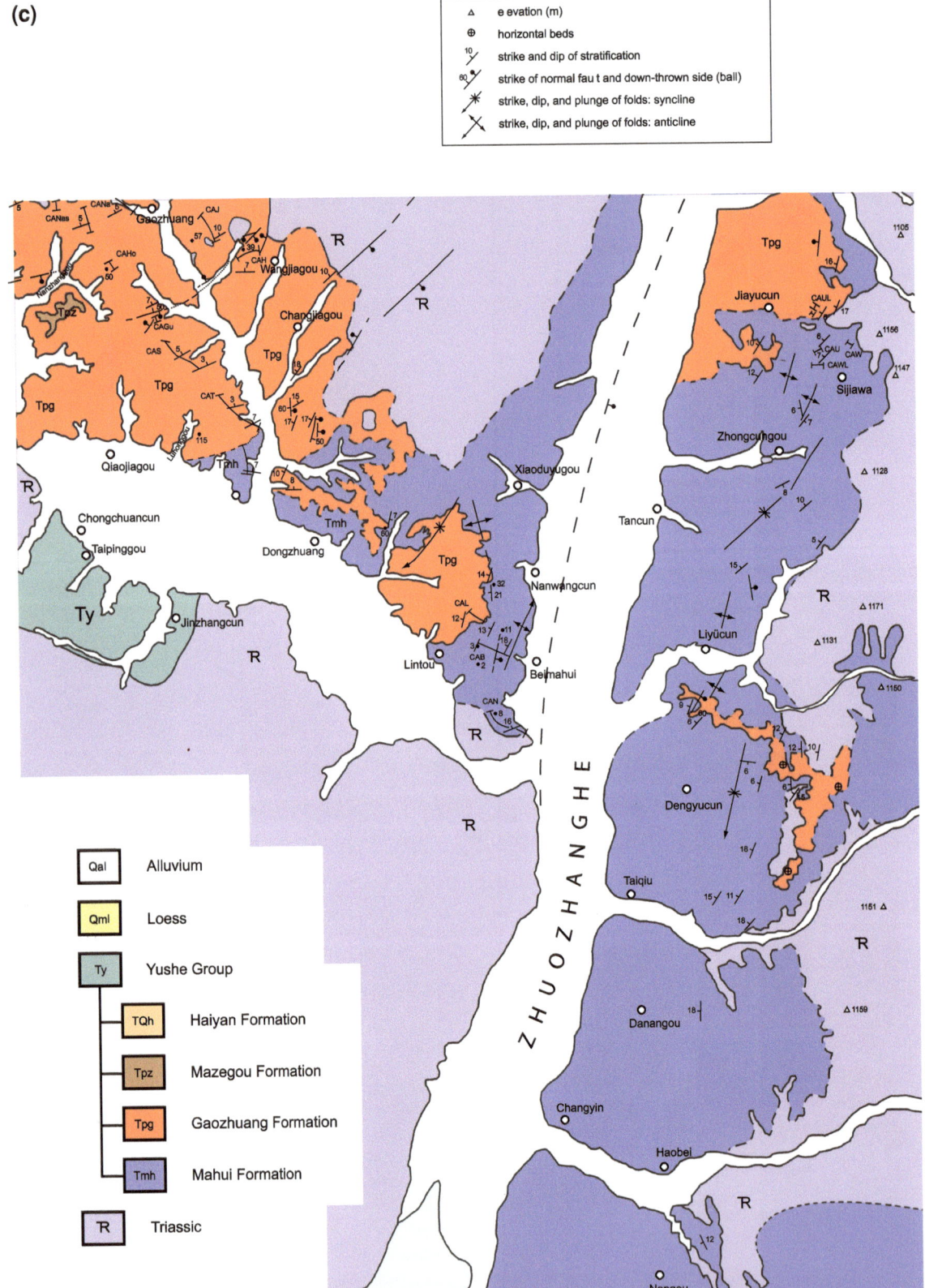

△	e evation (m)
⊕	horizontal beds
10	strike and dip of stratification
60	strike of normal fau t and down-thrown side (ball)
✳	strike, dip, and plunge of folds: syncline
✕	strike, dip, and plunge of folds: anticline

Qal	Alluvium
Qml	Loess
Ty	Yushe Group
TQh	Haiyan Formation
Tpz	Mazegou Formation
Tpg	Gaozhuang Formation
Tmh	Mahui Formation
℞	Triassic

0.5   0   1.0 km

Fig. 3.1   (continued)

◀Fig. 3.1 a–c Geological map of the northern Yushe subbasins prepared by the Sino-American mapping teams (1987–1998), showing lithostratigraphy, structure, and traverses used to construct the stratigraphic section and correlative magnetic sampling. Selected fossil localities shown by *black dots* and identifying YS numbers. Village names represented by *black circles* identify collecting points for historic collections. **a** focuses on the Yuncu Subbasin, the westernmost of the subbasins (outlined in Fig. 1.2), with the most complete section. **b** Geological map of the northern Yushe subbasins, continued. **b** shows lithostratigraphy, structure, and traverses used to construct the stratigraphic sections for the Nihe-Ouniwa area and northern part of the Tancun subbasin. Selected fossil localities shown by *black dots* and identifying YS numbers. **c** Geological map of the northern Yushe subbasins, continued. **c** includes most of the Tancun subbasin and lower Yuncu subbasin deposits on opposite sides of the Zhuozhang River, and shows lithostratigraphy, structure, and traverses used to construct the stratigraphic sections and magnetic sampling. Selected fossil localities shown by *black dots* and identifying YS numbers

Cenozoic deposition in this region of external drainage presumably involved disruption of gradient due to uplift along the axis of the Taihang Shan. Sediments began to accumulate in the paleovalleys during late Miocene time. As deposition proceeded, continued uplift along the Taihang Shan shifted the locus of deposition westward into the upper reaches of the paleovalley system. Thus not only does the entire basin-fill dip to the WNW with decreasing attitude, but the deposits become progressively younger to the northwest. This mechanism accounts for the presence of thick basin deposits (up to 800 m) accumulating in valleys with a maximum relief of only 700 m. There is little evidence that the deposits ever significantly overtopped the ancient interfluves to unite the basin fill, instead each branch of the original drainage maintained its own characteristic interplay of colluvial, fluvial, paludal and lacustrine facies depending on the size and nature of each catchment. Hence each paleovalley forms a subbasin with its own stratigraphy that limits easy development of a basin-wide lithostratigraphic synthesis. Historically chronostratigraphic schemes have been used to subdivide the deposits, but as study of the basin-fill has progressed, this has been replaced by lithostratigraphic criteria. Currently, basin-wide synthesis has become increasingly practicable because magnetostratigraphic studies in separate subbasins reveal local tectonic histories (deposition punctuated by hiatuses) and allow perception of structural and depositional events on a larger scale.

3.2 Previous Work

The first scientific notice of fossil mammals, and hence Neogene deposits, in the Yushe Basin was given by Geological Survey of China reconnaissance teams from 1918 to 1933. J. G. Andersson's success in China as mining advisor attracted the financial support of Axel Lagrelius and establishment of the Kinafond, which enabled Otto Zdansky to conduct fieldwork in Shanxi Province (Lucas 2001). Zdansky famously made the Baode *Hipparion* red clays of northern Shanxi known to western science, but he apparently did not spend time collecting in Wuxiang County. Some of the specimens purchased in Yushe and Wuxiang counties were described by Zdansky (1925a, b, 1927a, b) and others of the Swedish group assisting the Geological Survey.

In 1931 the Geological Survey's indefatigable collector-explorer Liu Xi-Gu also visited the region and reported the extensive presence of fossil remains. This prompted Pierre Teilhard de Chardin and C. C. Young (Yang Zhong-Jian) to begin an exploration of southeastern Shanxi. As they described their journey in 1933 they entered the Yushe Basin at Houmu (now Gengxiu) at the north and, passing southwestward through Yushe, Wuxiang and Qin counties, exited at the southeast corner of the basin at Hsintien (now Xindian). Confined to observations adjacent to the road, they did not get a complete impression of the lateral extent of the basin but its general stratigraphy and setting were accurately assessed. The Cenozoic deposits were seen to be contained within the axial part of a synclinorium developed in Triassic rocks and to dip westward more or less in conformation with the eastern limb of the Triassic fold, so that the base of the Cenozoic strata was exposed only along the eastern border of the basin.

In the reconstruction of Teilhard de Chardin and Young (1933, Fig. 2.8 herein) the basin fill includes three stratigraphic units. (1) At its base, the "torrential-lacustrine series" began with a basal conglomerate of rounded Triassic boulders succeeded by violet sands and clays and yellow sands. (2) The "red loam", concretionary clay-rich sediments banded in darker red (paleosols), was distributed in accordance with the present drainage system, lying unconformably on a deeply dissected surface cut into all older rocks. At Zhangjiagou (=Zhangcungou), Wuxiang County, the red loam was divisible into two units, the older of which was darker red, horizontally bedded and disconformably overlain by the younger lighter colored unit. (3) The "loess", separate from the red loam, lies on a deeply dissected surface, mantling the present landscape but now itself considerably dissected. Fossils collected by the party at Houmu showed that the lower part of the "torrential lacustrine series" was of "Lower Pliocene, or Pontian" age. Scattered finds made by Liu Xi-Gu also indicated that deposits of "Sanmenian" age, regarded as late Pliocene, may lie in the upper part of this "series". A large deer antler fragment [*Cervus (Elaphurus)* cf. *bifurcatus* Young, 1935] from the older "Red loam" suggested these were also of "Sanmenian" age, while the overlying "red loam" (lower case "r") contained remains of the living molerat or zokor, *Eospalax fontanieri*, and thus were no older than the "Choukoutian stage" or early Pleistocene.

Two years later the Jesuits Licent and Trassaert (1935) spent 45 days during July and August of 1934 supervising excavations and purchasing material from country people in the "Yünchuchen Basin" (now Yuncu Subbasin) of Yushe County and the adjacent smaller "Changtsun Basin" (Zhangcun Subbasin) of Yushe and Wuxiang counties. From this work a tentative biostratigraphy arose which was generalized by these authors as three chronostratigraphic units. These units assume considerable importance, for despite ambiguities between text, map, sections and faunal lists, this work remained, for 40 years, the sole basis for the stratigraphy of the Yushe Basin. The original stratigraphic basis for the threefold division of the "Pliocene-lacustrine series" of Licent and Trassaert (1935) grew from observations on the following sections.

Zone 1 was "mostly exposed and fossiliferous near *Lingt'ou*" (Lintou) in the eastern Yuncu subbasin and represented by "hard consolidated conglomerates and dark red sandstone, immediately derived from the underlying Permo-Triassic beds." Zone 1 contained a "typical Pontian fauna" (Licent and Trassaert 1935).

Zone 2 was most completely studied in the Zhangcun subbasin. Compared with the Zone 1 deposits, it is "less coarse, and a typical lacustrine condition is prevailing: green and bluish marls, containing many bird, turtle, fish-remains, fresh water shells… and plant remains" (Licent and Trassaert 1935). These features and their superpositional relationships were detailed in a 60 m measured section (ibid, their Fig. 1.2) near *Changtsunkou* (Zhangcungou). Zones 1 and 2 "are possibly continuous being chiefly distinguishable lithologically by an increasing lacustrine condition." "A middle Pliocene age seems to be indicated" by the mammalian fauna.

Zone 3, mostly studied in the Yuncu subbasin, was "chiefly sandy, with the presence, however, of two layers of marl at the middle of the deposits, those layers indicating perhaps a maximum in the lacustrine conditions. In that case the zone would correspond to a complete sedimentary cycle by itself" (Licent and Trassaert 1935). It rests with "clear erosional breaks and even overlapping" upon Zone 2. This zone contains "the appearance of the *Equus* fauna" correlative with that from Nihewan then regarded as of Sanmenian age and previously considered late Pliocene.

Licent and Trassaert realized that this sequence (Fig. 3.2) may be the most complete representation of deposits of Pliocene age in eastern Asia. When Teilhard de Chardin and Trassaert began to monograph the accumulated fossil collections in 1937 they were particularly interested in the nature of the fauna of Zone 2, which they realized contained a hitherto unknown phase in the history of Asia's mammals. Teilhard de Chardin and Trassaert (1937a, b) signalled a subtle change in emphasis toward faunal characterization of the Yushe zones by relabelling them in

Fig. 3.2 Cliff of thin-bedded sandstone, siltstone and claystone capped by marl. Zhan-Xiang Qiu (*red backpack*) and Jie Ye standing on Triassic outcrop at base of Licent and Trassaert (1935) section (see Fig. 2.12). Lin-Zhong Hao sits on midslope at approximate level of fossil occurrence. The *top* of this outcrop is in the near distance

roman numerals. Much of the collection had been purchased from the county villages with little locality data beyond the name of the village purchase point. This necessitated allocation to zone using knowledge of "early Pliocene" faunas, principally from sites near Baode in northwestern Shanxi, and the late Pliocene and younger faunas of Nihewan of northern Hebei as well as stage-in-evolution criteria to deduce the nature of the middle Pliocene assemblage. Only the excavations in Zhangcungou provided objective criteria for the association of some elements of this fauna.

After 1937 the Yushe and Wuxiang region became part of the battle-zone prohibiting work there. Teilhard de Chardin continued publishing on fauna from Yushe and elsewhere (Teilhard de Chardin 1942; Teilhard de Chardin and Leroy 1945a, b). A review of Cenozoic geology of the middle and lower Huanghe by Pei et al. (1963) transformed Teilhard's "Yushe series" to the Yushe Formation and relegated the zones to lower, middle and upper divisions, but no further conceptual advance was made. New geological investigations were deferred until the early 1980s when the Geological Bureau of Shanxi Province began its 1:200,000 scale mapping of the province. The Yushe Basin was contained in the Pingyao Quadrangle (Geological Bureau of Shanxi Province 1976) and the Yushe Formation was elevated to group status. The Pingyao Quadrangle (see Fig. 3.1) contained the best known sections in Yushe, Wuxiang and Qin counties; and the investigators chose the sequence in the Zhangcun Subbasin, lying just south of the Yushe–Wuxiang county line, as the type section for a three-fold subdivision of the Yushe Group. This stratigraphy was extended throughout the quadrangle. The lithostratigraphic units defined owed much to concepts of the 1930s: the Renjianao Formation represented the coarse colluvium and alluvium at the base of the section; succeeded by the Zhangcun Formation, mostly fine sands and clays of a dominantly lacustrine nature; and at the top, the Louzeyu Formation, a sequence of sands, clays and marls representing a mostly fluviatile regime with minor lacustrine events. The latter unit was inferred to rest unconformably on the Zhangcun Formation, and cross-sections of the subbasin (Figs. 2.12, 2.23) show it truncating the upper units of the latter. As measured in the Zhangcun subbasin the Yushe Group was about 460 m thick.

In the southeastern part of the basin, covered by the Qinyuan Quadrangle, the sequence is thinner, 70–150 m and restricted to small basins inset in the Triassic terrain. A local stratigraphy was proposed: a lower unit, the Siting Formation of fluviatile sands and gravels, is overlain by the Beiji Formation of finer sands, clays and marls. This local stratigraphy was subsumed into the Renjianao and Zhangcun formations in later syntheses (e.g., Cao et al. 1985) because of gross similarity in lithology and sequence. Thus the Yushe Group stratigraphy, proposed for the Pingyao Quadrangle, has become widely accepted by later workers. The Renjianao and Zhangcun formations were relegated to the Pliocene and the Louzeyu Formation to the early Pleistocene.

Some resolution of the relationships of the overlying loessic deposits was also proposed by the Shanxi geologists in accordance with concurrent syntheses of loess stratigraphy over the whole of the loess plateau. The younger reddish loam (lower case r of Teilhard de Chardin and Young 1933) was equated with the widespread Lishi Loess and the younger blanketing yellow loess with the Malan Loess. The nature and relationships of the older red loam (capital R of Teilhard de Chardin and Young 1933) was a further problem. The Shanxi geologists mapping the Qinyun Quadrangle termed such deposits the Daqiang Formation and that term has been widely used throughout the Yushe Basin.

In the late 1970s and the 1980s new studies of the stratigraphy, sedimentology, paleoenvironment, paleontology and magnetostratigraphy were initiated principally in the Zhangcun and Yuncu subbasins by two groups of workers. Cao Jia-Xin of Beijing University, her students and coworkers (Cao et al. 1985) concentrated on the Zhangcun subbasin where a well exposed section, rich in a variety of fossil organisms, was used to reconstruct the paleoenvironments of this area. The stratigraphy followed that proposed by the Shanxi geologists with the implication that major events in the Zhangcun subbasin held for the whole of the Yushe Basin. A lacustrine hypothesis for the basin was proposed in which initial fluviatile environments (Renjianao Formation) give way to lacustrine (Zhangcun Formation) followed by lake regression and fluviatile replacement (Louzeyu Fm.). The palynological record, in particular, tracked these sedimentological changes in the Zhangcun subbasin showing mesic floras of the lacustrine phase giving way to more xeric assemblages in the final fluviatile environments.

Studies in the contiguous Yuncu and Tancun subbasins by Qiu Zhan-Xiang and colleagues at IVPP and the Tianjin Museum showed that the stratigraphic succession there is more complex and dominated by fluviatile environments. The Yushe Group is nearly twice as thick in this area, which is separated from the Zhangcun subbasin on the south by ancient interfluves of Triassic rocks. It was thus difficult to apply the lithostratigraphy of the Zhangcun subbasin to that observed in the adjacent subbasins. Qiu et al. (1987) resolved the lithostratigraphic sequence in the Yuncu subbasin into four sedimentary cycles each of which begins with coarse clastics and fines upward to lacustrine or paludal clays and finally marls. Accordingly, a new lithostratigraphy was proposed. The Mahui Formation, containing a basal colluvium and characterized by yellow cross-bedded sands and local cobble-boulder conglomerates fines upward to muddy sands, clays and finally marl. It is locally overlain unconformably by the Gaozhuang Formation dominated by fluviatile sands and conglomerates ending in mudstones and marls, which are disconformably overlain by the Mazegou Formation, hard muddy sandstones and mudstones, the top obscured by alluvium. The final unit, the Haiyan Formation, is flat-lying and lies with angular unconformity across the gently dipping Mazegou Formation. It is composed mostly of fine sandstones and siltstones with local clays and marls.

In 1991 Bao-Yu Huang and Shu-Yuan Guo again summarized the stratigraphy of the "Yushe–Wuxiang Region" (Huang and Guo 1991), reiterating much of the work of the Shanxi geologists (1975–1976) except to recommend subsuming the Louzeyu Formation into the Zhangcun Formation on the basis that the "type cross-section" of the Shanxi

Fig. 3.3 Looking northeast across the head of the Zhangcungou to the slopes of loess-mantelled Triassic sediments that form an interfluve between the Zhangcun and Yuncu subbasins. Basin fill in the near distance is the *top* of Licent and Trassaert (1935) section held-up by marl

geologists, measured near Zhangcungou in the Zhangcun subbasin, also included Licent and Trassaert's (1935, their Fig. 1; see our Figs. 3.2, 3.3) 29 m type section for Zone 2. Subsequent field work and our own observations (Sect. 3.7.2) reinstate the Louzeyu Formation as a distinct body of sediments.

Huang and Guo's (1991) contribution includes a useful facies analysis delimiting the depositional environments of the Yushe Group based mainly on the Zhangcun subbasin but with some remarks on conditions in other subbasins to the north. Both Huang and Guo (1991), and Wu (1991) in her chapter in the same volume review the tectonic history of the region prior to deposition of the Yushe Group and comment, as we have above, on the syntectonic effects during deposition of the unit. They also give a synthesis of the depositional history, again derived mainly from observations in the Zhangcun subbasin, which proposes a lacustrine model of waxing and waning of a basin-wide lake system. Wu (1991) provides a grain-size analysis to characterize sedimentary environments along with mineralogy, and clay mineral and chemical analyses to assess water salinity and paleoclimate for the upper part of the Zhangcun

Formation. She comments on the fish, macro-plant remains and pollen for the same part of the section to indicate a trend to greater salinity and xeric environments at the close of Zhangcun deposition.

The most recent study of the Zhangcun subbasin is by Shi Ning (1994) whose review of the evidence from that subbasin is included in a subsequent section of this chapter. Among his many contributions to geology, chronology and paleoenvironment of that subbasin, his review of the lithostratigraphy, based on a new measured section, led him to break the Zhangcun Formation of other authors into two units, a lower Wangning Formation (new) and an upper Zhangcun Formation (revised), and to reiterate the validity of the Louzeyu Formation and provide a revised definition of its base. He advocated a return to strictly lithostratigraphic base for rock units and the recognition of unconformity-bounded sedimentary cycles or natural groupings of strata. Although not explicitly stated, the cycles consist of major fining-upward units beginning with thick conglomeratic sands and extending through mixed sand, mudstone and clay lithologies to clays and terminal marls (Fig. 3.3). This style of vertical facies changes can be seen at smaller scales within the major

(a)

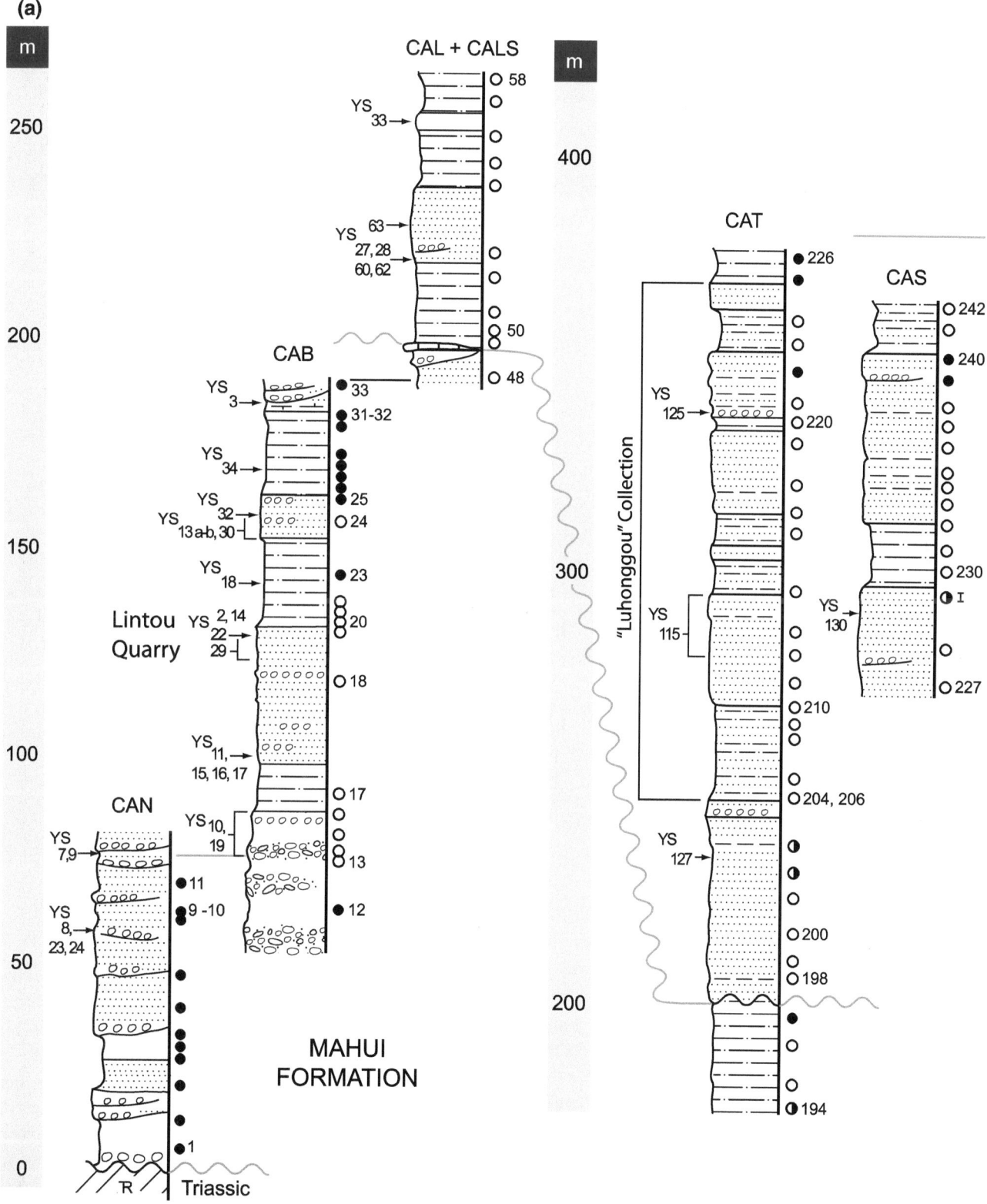

Fig. 3.4 a–d Measured sections for the Yuncu subbasin with correlative paleomagnetic sampling sites (filled and *open circles* denoting normal and reversed polarity, respectively, numbers to *right* identify the samples) for determination of Virtual Geomagnetic Poles (VGP). YS field locality numbers on *left* are fossil sites. **a** sections are stratigraphically lowest, including the Mahui Formation and, above the unconformity (*wavy line*), the basal Taoyang Member of the Gaozhuang Formation. **b**, as for **a**, is a set of sections through the Nanzhuanggou Member of the Gaozhuang Formation. **c**, as for **a**, includes the Culiugou Member as well as part of the Mazegou Formation. **d**, as for **a**, includes upper portions of the Mazegou Formation with the overlying Haiyan Formation

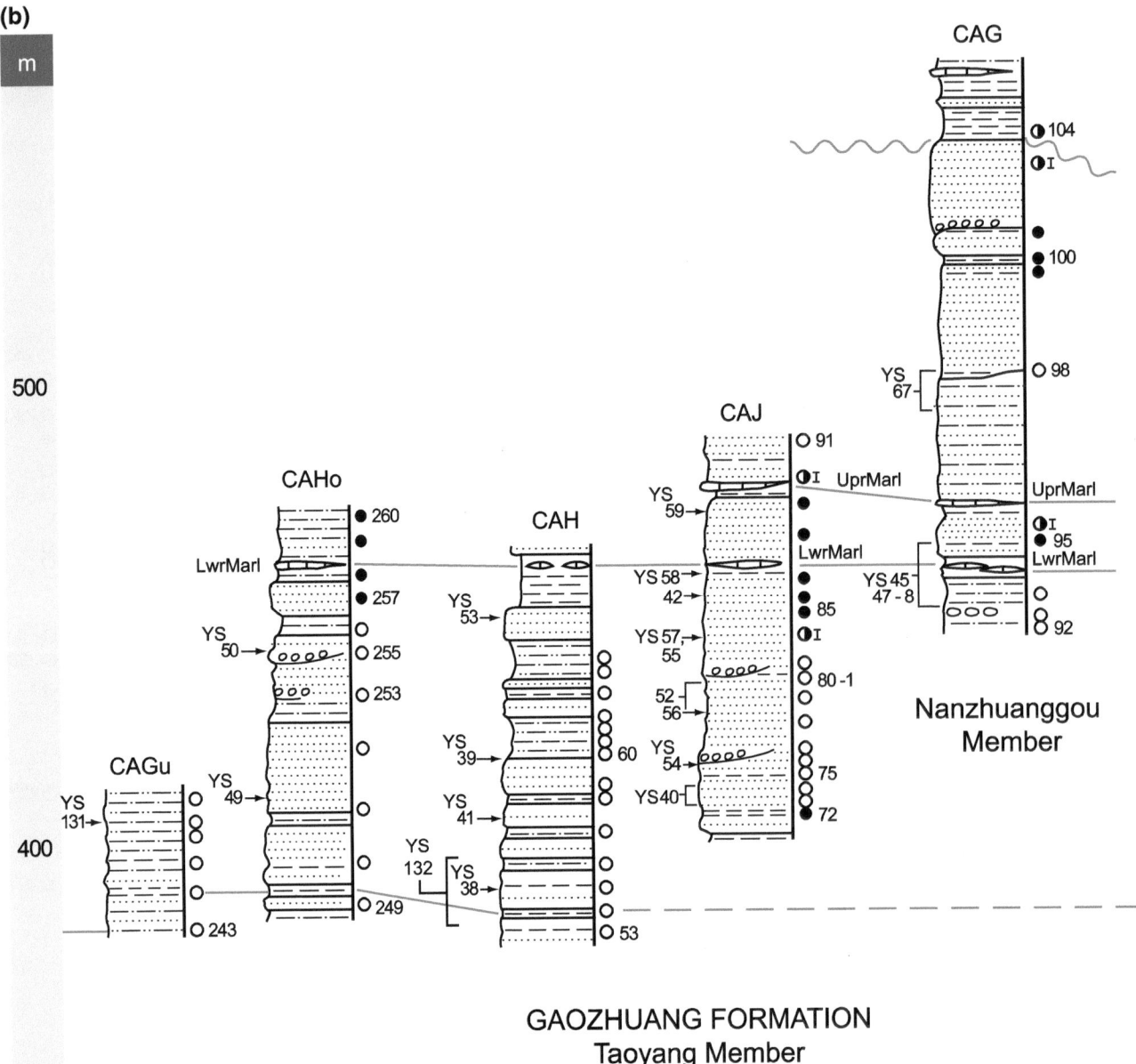

Fig. 3.4 (continued)

cycles as well, but these do not define subbasin-wide hiatuses. Adopting these principles led Shi to revise the subdivision of the stratigraphic column in the Zhangcun subbasin and thus to different estimates of thicknesses for units than those determined by the Geological Bureau of Shanxi Province (1976). He found no appreciable hiatuses at the unconformities bounding the major cycles in contrast to our work in the Yuncu subbasin.

3.3 Methods

We used the well-known methods of field geology in our studies but did not follow-up with laboratory investigations of the sort that were used in previous studies of the Zhangcun subbasin. Grain size texture was estimated with reference to comparators calibrated to the scale graduated in millimeters rather than the commonly used phi-scale. Rock color was

(c)

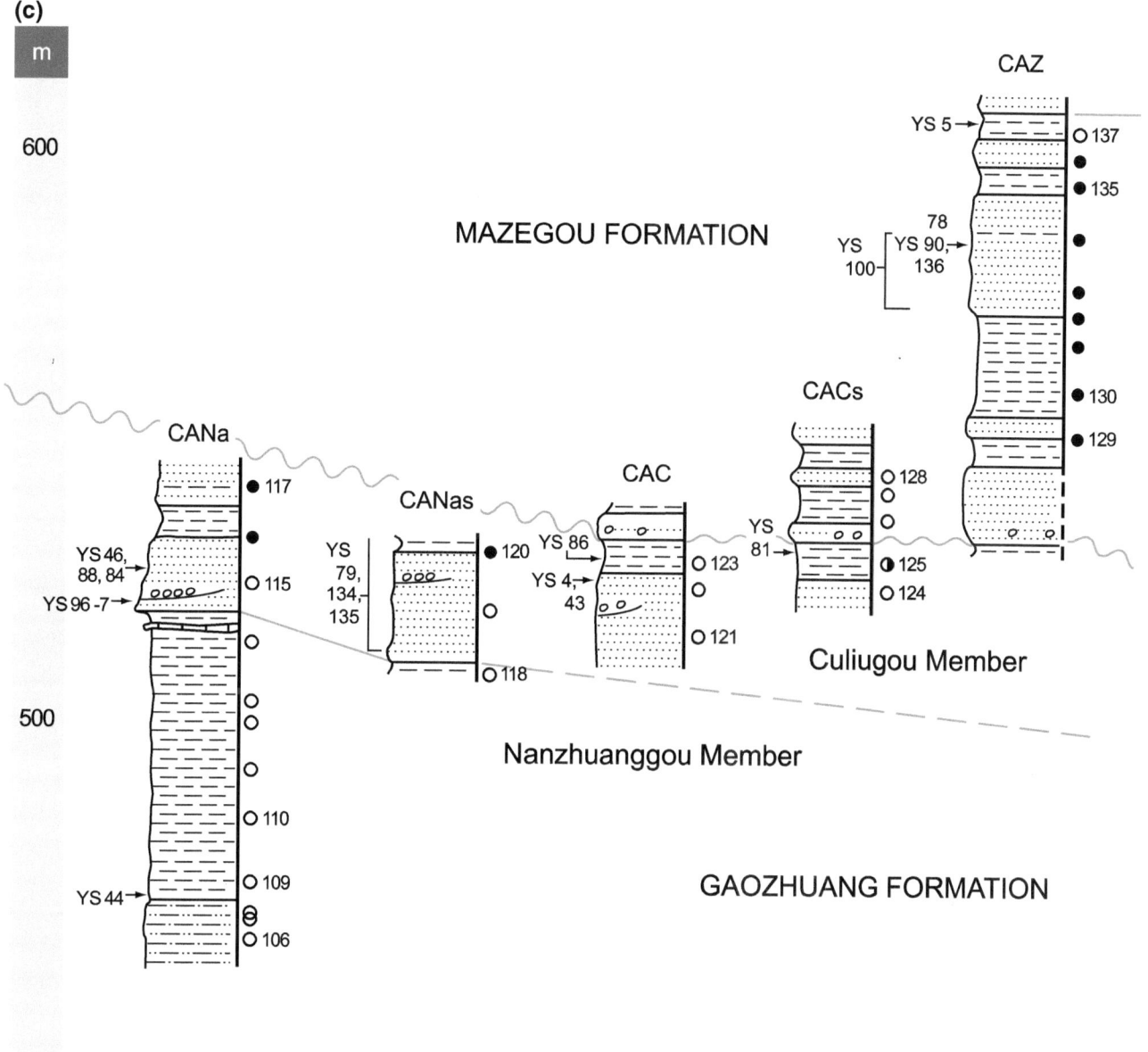

Fig. 3.4 (continued)

quantified with reference to Munsell charts (Geological Society of America, 1975) and sedimentary structures are described in the usual terms (Tucker 1982). Stratigraphic thickness was measured using a special 1.5 m Jacob staff designed by one of us (Ye Jie), based on his long practical experience in field methods. Simultaneously with section demarcation and measuring, paleomagnetic samples were taken under the supervision of Neil Opdyke, University of Florida. Paleomagnetic sites were placed precisely into the section, as were fossil localities identified during our campaign.

Because of the ubiquitous and thick cover of loess, the total section of Yushe Group rocks was pieced together from successive subsections correlated by traceable beds. These correlations were later tested by the magnetostratigraphy that

followed the same traverse. In graphic presentation (Fig. 3.4a–d) the lithological log is generalized from the original data to a reduced scale in which only units of thickness of 2 m or more are shown. The positions of fossil localities and magnetic sample sites are not distorted; the latter were collected at 3 m or more intervals depending on lithology.

The stratigraphic column was divided into mappable units following the principles enunciated in Qiu et al. (1987) and Shi (1994), which recognize natural unconformity-bounded units, each embodying a major fining-upward sedimentary cycle. These units were mapped on 1:25,000 enlargements of the 1:50,000 Yushe and Wuxiang quadrangles and reduced to the latter scale in compilation (Fig. 3.1). Contacts of lithostratigraphic units and structural symbols follow well established conventions.

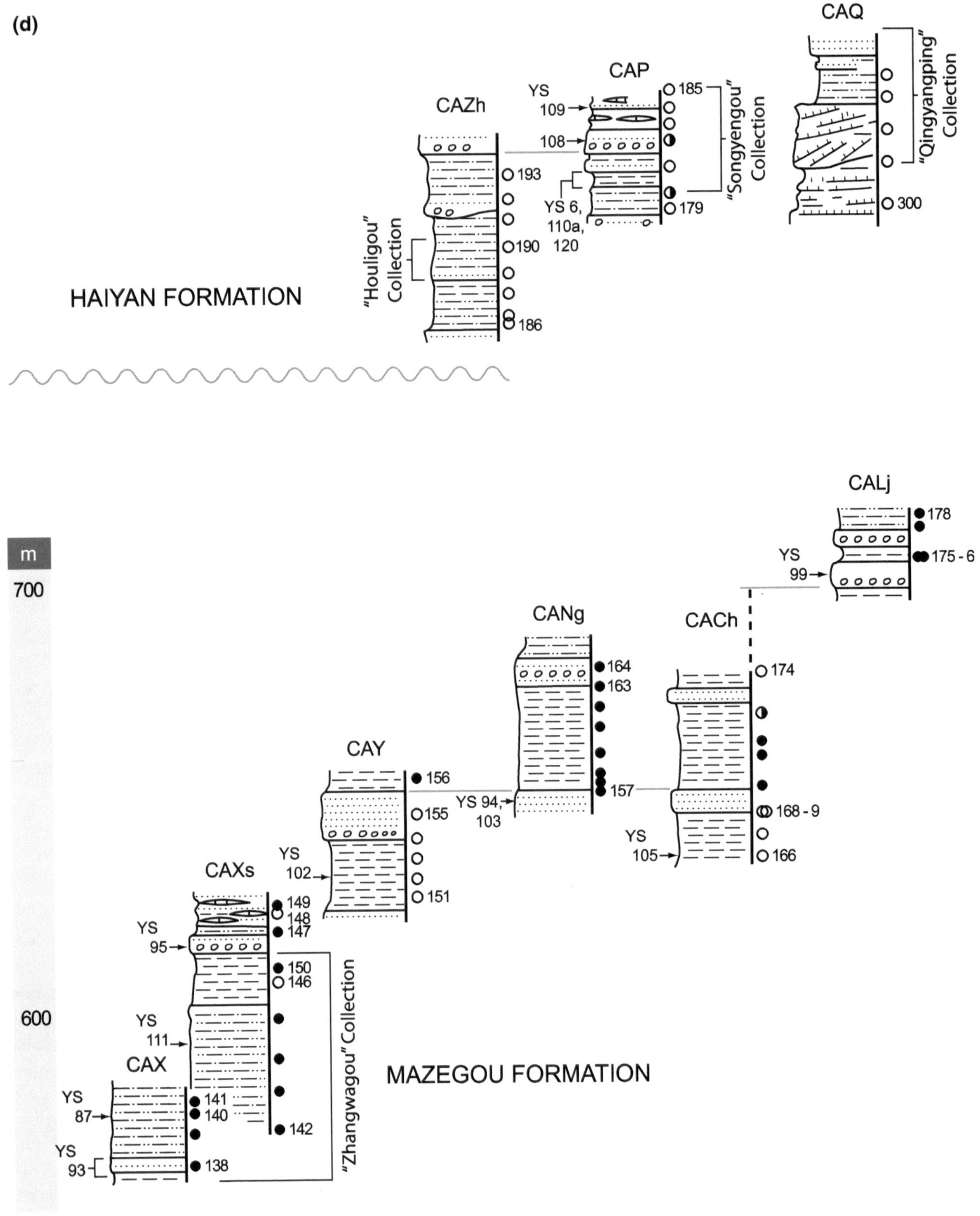

(d)

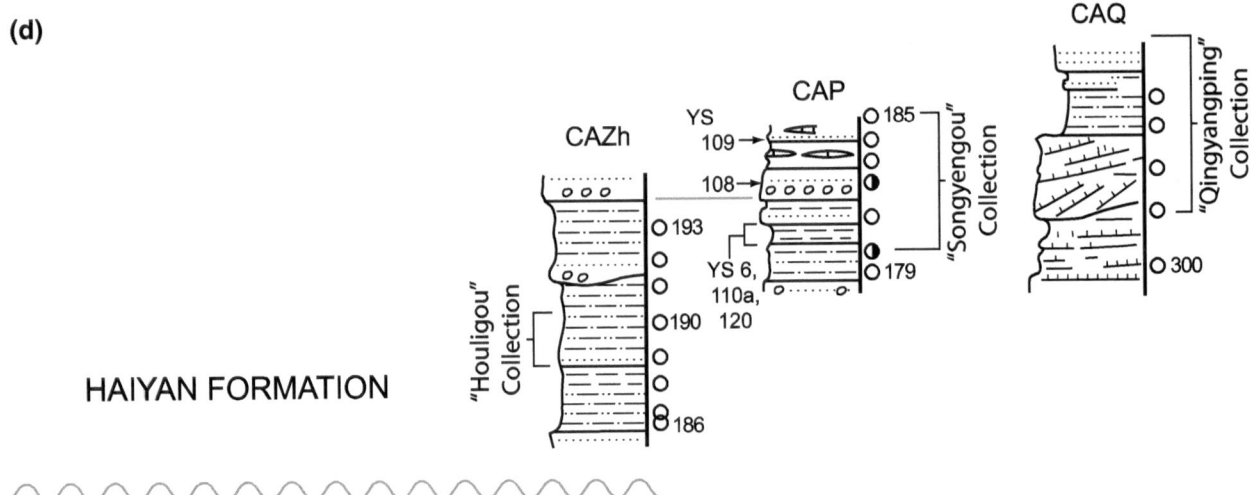

HAIYAN FORMATION

MAZEGOU FORMATION

Fig. 3.4 (continued)

3.4 Pre-Yushe Group Events

The Yuncu and Tancun subbasins are broadly interconnected in the northern part of the paleovalley system that constitutes the Yushe Basin (Fig. 1.2). The Tancun subbasin is the eastern part containing early deposits of the Zhuozhang River. It is connected along that river with the Ouniwa subbasin to the north. The Nihe subbasin seems to be broadly connected with the Ouniwa subbasin, lying in a western tributary comparable to the Yuncu subbasin. The Yuncu subbasin is a major western tributary whose older deposits are extensions of strata from the Tancun subbasin and whose younger rocks are restricted to the western part of its drainage.

Triassic rocks beneath the Yushe Group belong to the continental Ermaying Formation, which consists of a repetitive succession of thick to thin-bedded, fine to medium, greenish-gray sandstones and interbedded red-violet claystone in broadly lenticular bodies. These continental rocks form the dominant source of sediments for the Yushe Group. A minor, yet sometimes conspicuous, component of quartzose rocks was contributed from the limbs of the synclinorium where Paleozoic rocks were exposed on the drainage divides of the Yushe Basin. Thus well rounded quartz, quartzite, chert and rare jasper, mostly as pebbles and no larger than cobbles, are often present in most Yushe Group conglomerates in addition to the dominant Triassic sandstones.

Prior to Miocene fluviatile aggradation, the region encompassed by the Yushe Basin had been reduced to the topography of low relief into which deep dentritic drainage was imposed. Interfluve crests have subdued topography marked by low rises to maximum elevation. Valley wall slopes varied from moderate to steep, with the steepest slopes bearing on a north-northeast aligned fault and joint pattern that guided local drainage incision. Exhumed basement slopes beneath the Yushe Group in the Yuncu and Tancun subbasins bear only a thin cover of regolith. Large and small blocks of Triassic sandstone, rounded to a degree, rest on such slopes engulfed in younger sediments. These may have been core-stones within the weathered mantle and show little sign of transport. Stripped basement surfaces are weathered smooth and fractured, the open fissures filled with calcite cemented fine regolith.

3.5 Yushe Group Stratigraphy in the Yuncu and Tancun Subbasins

3.5.1 Mahui Formation

Description. Defined by Qiu et al. (1987), this unit represents the first major Tertiary sedimentary cycle preserved in the Yuncu and Tancun subbasins. The unit is coarser than suceeding formations, its thickness being about 50 % conglomerate and sandstone. The remaining thickness is mostly mudstone and minor clays and marls. Like all the Yushe Group sediments, the Triassic basement is the major source of the Mahui Formation and thus the range of lithologies and colors is monotonously similar throughout.

The type section of the Mahui Formation lies in the easternmost Yuncu subbasin. It extends from the contact with basement just north of Nanmahui, continues north to near Beimahui and turns northwest up the canyon west of the latter village to the top of the outcrop exposed beneath the loess on the divide between the Zhuozhang and Yuncu rivers (the CAN and CAB sections, see below). About 200 m of sediments are exposed in this traverse. There is only 4 m of colluvium at the base, represented by a breccia of angular Triassic blocks from pebble to boulder size (up to 70 cm) in a fine to medium clayey red brown (10R5/4) muddy sand matrix. This is overlain by stream-reworked talus of muddy pebbly sands followed by buff (5YR7/2) poorly-sorted sands with lenses of pebble to boulder conglomerate composed mostly of Triassic clasts, but with some pebbles and granules of quartz and quartzite indicating sources at the periphery of the drainage system. A thick interval (25 m) of conglomerate occurs in the type section beginning about 55 m from the base. It is composed of lenses of rounded pebble to boulders of Triassic sandstones with minor pebbles of quartz. Imbrication of the flatter clasts indicates transport to the south and southwest. These conglomerates pinch out westward, where they are replaced by large-scale (1–2 m sets), festoon cross-laminated, yellow (10YR7/6) fine to coarse sands with lenses and clasts of mudstone. Cross-laminated yellow friable fine to medium sand with lenses of pebble conglomerate continue upward to 130 m from the base where the first major interval of mudstone begins. Mudstones, typically violet (5R6/2-10R4/4) in color, become important units in the upper part of the Mahui Formation, and are interbedded with yellow sands and pebbly sands. In the uppermost 10 m of the type section (top of CAB and base of CAL measured sections) the unit is a yellow (10YR6/4) laminated siltstone, yellow and variegated calcareous siltstone and claystone with irregular calcareous concretions (i.e., marl) capped by white flaggy limestone that forms a prominent dip slope that is laterally traceable for some distance. Local channel fills of pebbly medium sand beneath the capping limestone cut up to 5 m into the underlying marly interval and contain pebbles of the calcareous concretions within the marl.

The Mahui Formation can be traced westward into the Yuncu subbasin beneath the Gaozhuang gravels west of Dongzhuang village to the mouth of the Gaozhuang valley. At that point it is truncated by a zone of faulting manifested by nearby fault trends, steepening dips and changes of strike. In the CAT subsection, north of Taoyang, the basal

27 m of interbedded sandstone and mudstone are recognized as the Mahui Formation (principally due to a normal paleomagnetic sample) beneath the basal Gaozhuang sandstone. Yuncu subbasin sections spanning the Mahui, Gaozhuang and higher formations are aligned in Fig. 3.4; see especially Fig. 3.4a.

The Mahui Formation crops out all along the valley of the trunk stream that is occupied today by the Zhuozhang River, and continues along the eastern side of that valley in the Tancun subbasin where it rests on steep slopes of Triassic sandstones and mudstones, and fills a dissected topography usually without a basal colluvium. Only lenses of talus breccia are interbedded with the finer fluviatile sediments and in many places sands and even thin marls rest directly on the weathered surface of the Triassic. This Mahui paleoriver is represented by boulder conglomerates filling a paleo-canyon communicating with the Ouniwa subbasin to the north. To the south it apears to be represented by a similar canyon-filling coarse facies just west of the Guanhe Reservoir in the vicinity of Zhangcungou (the suffix "-gou" means valley or gully) where it enters the Wuxiang Subbasin as observed by Teilhard de Chardin and Young (1933) in 1932. One of us (Qiu) has observed similar conglomerates along the western side of the Tancun subbasin south of the mouth of the Yuncu River. North of that point such conglomerates appear in the lower part of the Mahui type section. Where clast imbrication can be seen, the transport direction is southerly. Lateral tracing in the Beimahui area shows that these coarse deposits rapidly pinch out westward into festoon cross-bedded sands that give way to interbedded sands, muds and marls as the most distal facies.

Our most intensive study of the Tancun subbasin was confined to its northern end just south of the Yushe County seat. There a measured section begins just north of Sijiawa village and steps northwestward in a series of subsections to the highest point on the ridge on the north side of the valley containing Jiayucun ("cun" meaning village). Only 130 m of section was present in this traverse which begins a few meters above the basement (Fig. 3.5). Of this thickness only 80 m represents the Mahui Formation. In this area Mahui sediments are mostly sandstone and pebbly sandstone, buff (5YR6/3) in color, interbedded with thinner violet (10R6/2) mudstones and two widely traceable marls. The marls are white flaggy limestone or nodular limestone, often in multiple thin beds, 30–40 cm in thickness, capped or interbedded with light green (5Y7/3) and yellow-mottled claystone. The lower marl is 15 m from the local base of the section (CAW subsection) and the upper marl lies 33 m higher (recorded together in the CAW subsection). Laterally both marls may be removed by local disconformities at the base of overlying pebbly sand bodies. These marls can be traced southward to Liyücun 7 km south of Yushe town, but to the north they cannot be surely correlated with the thin marls in the northeastern corner of the subbasin. Likewise we were not able to carry the marls

southward into the outcrops north of Dengyucun. Their northeast strike indicates that they should pass southwest into loess covered terrain.

Geologic mapping from east of Dengyucun to south of Haobei village, 16 km south of Yushe, indicates an additional thickness of Mahui Formation and some complexity in its stratigraphy. The lack of marker beds made reconstruction of the composite section difficult, but it was possible to assemble parts of the sequence and get an approximation of the composite thickness.

Between Taiqiu and Haobei villages boulder conglomerate rests on an irregular surface of several meters relief cut on the Triassic (Fig. 3.6). Little colluvium was seen on the Triassic bedrock, except for coarse river-born boulder beds. These deposits crop out as a sheet of conglomerate up to 20 m thick that blankets the Triassic in a broad arc convex eastward between 2 km east of Taiqiu and a kilometer east of Haobei. The boulder beds form the base of a sequence of sands and mudstones lying in a deep valley northeast of Taiqiu that extends eastward into the adjacent quadrangle. A partially buried ridge of folded Triassic rocks extends SSW from the basement mass east of Liyücun to the river east of Taiqiu.

The boulder conglometate 2 km east of Taiqiu consists of two sheets separated by pebbly sands. The conglomerate is composed mostly of pebbles to boulders of Triassic sandstone, some flat and angular enough to reveal imbrication that indicates southward and southwestward flowing currents. This conglomerate contains conspicuous and relatively numerous pebbles of white quartz, gray quartzite and black chert, presumably from early Paleozoic and Precambrian rocks of the Taihang Shan to the east along with pink siltstone and red mudstone clasts derived from the nearby Triassic. These conglomerates are succeeded by pebbly sandstone and violet mudstone, the whole sequence about 30–40 m in thickness. Eastward in the valley, lateral equivalents of this conglomeratic interval are massive to thick bedded pink to light violet sandstones and pebbly sandstones with flaggy concretions interbedded with thin violet mudstones, more than 40 m in thickness. We did not have the opportunity to trace these rocks eastward.

Westward the conglomeratic sequence is overlain, in some places with apparent angular unconformity, but in others in apparent conformity, by the basal pebbly sandstones of a finer grained sequence that extends westward to the limit of outcrop. This latter unit comprises one major fining upward cycle at least 80 m in thickness that lies beneath the upper part of the Mahui contained in the measured section that we traced southward to Liyücun. It was difficult to establish continuity of these belts of outcrop due to gentle folding and considerable slumping around Dengyucun but there is no evidence that these sequences are not part of a single depositional cycle, the measured section

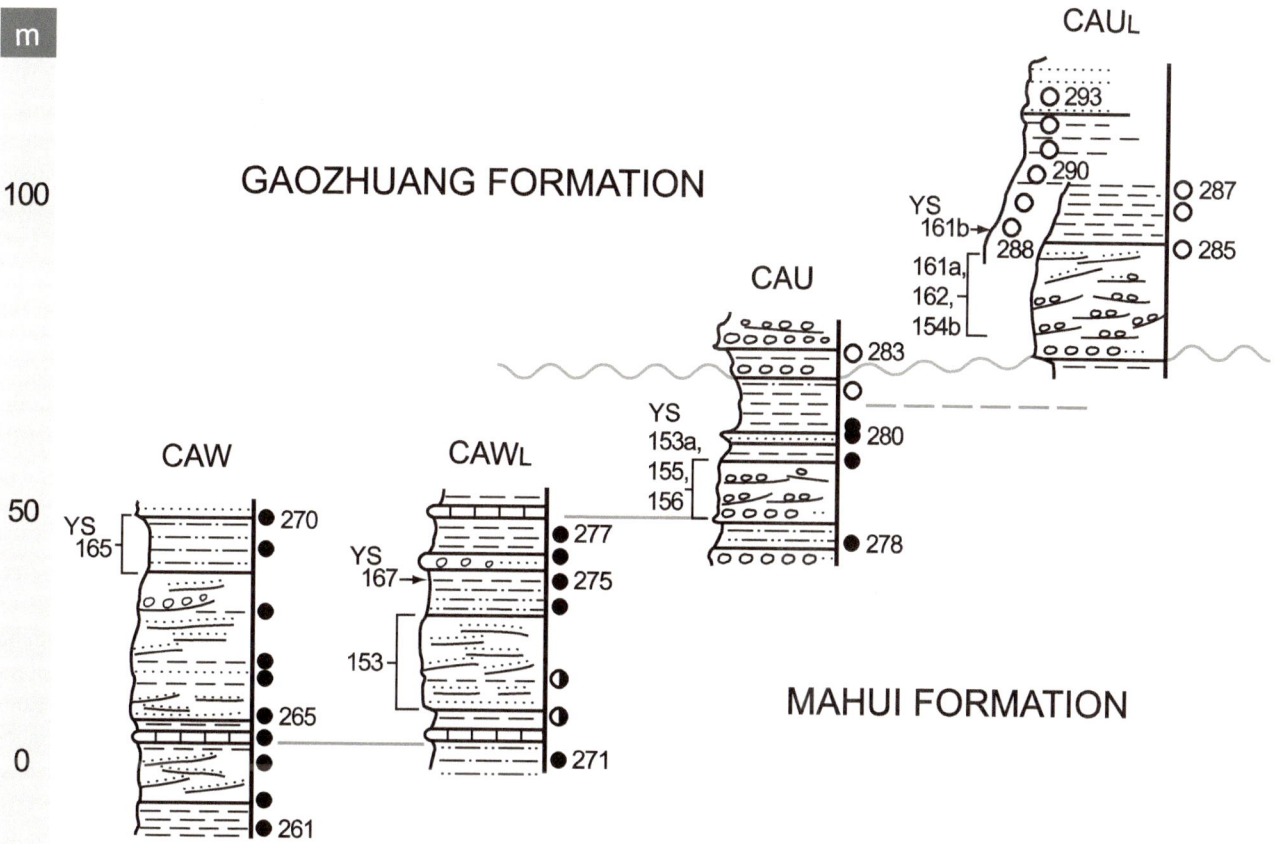

Fig. 3.5 Correlated measured sections for the northern Tancun subbasin (see Fig. 3.1b, c) showing sampling sites used to determine the sequence of Virtual Geomagnetic Poles (VGP). The hundred meters indicate dominantly normally magnetized sediment (Chron 3An) overlain by reversely magnetized deposits. The unconformity observed in the field (*wavy line*) is contained in the reversed magnetozone and probably represents relatively little time

Fig. 3.6 Looking northeast about 3 km east of Taiqiu, southern Tancun subbasin. Xiaoming Wang, Wenyu Wu, and Jie Ye walk along the contact at base of Yushe Group

representing the upper fine-grained part of the cycle (more clearly expressed distal to basement outcrop, e.g., compare CAW, near basement, with CAWL). If these conclusions are correct the Mahui Formation is at least 160 m thick in the Tancun subbasin, close to the measured thickness of 200 m in the Yuncu subbasin, both surely minimal estimates.

Geologic mapping and reconnaissance stratigraphic work north and northeast of Yushe town delineated the northeastern end of the Tancun subbasin. Triassic basement on the eastern flank of this area is overlapped passively by relatively fine-grained sediments. Even marls lap onto the steep Triassic slopes which, for the most part, lack colluvium. The sedimentary sequence includes a marl low in the section followed by yellow and buff sands, pebbly sands and interbedded violet mudstone that cannot be lithologically correlated in detail with the measured section south of Yushe. However, only a single cycle of deposition seems indicated and this can be correlated in general with the Mahui Formation. In the narrow valley north of Liutan the basal Mahui has a 20 m thick boulder conglomerate at the base that engulfs talus slopes in this embayment. Buff fine sands and thin platy limestones overlie these conglomerates. Gentle folds deform the sedimentary sequence so that the total section probably does not exceed 80 m in thickness.

The northwestern part of the outcrop is dominated by spectacular sheets of boulder conglomerate that fill a 3 km wide channel cut through the Triassic. These conglomerates are about 100 m thick, composed of rounded Triassic sandstone blocks from cobbles to large boulders and subrounded corestones from adjacent Triassic terrain up to 3 m in longest dimension. This conglomerate body is clast-supported with little sand matrix. Lenses of sand and even violet mudstone occur within these bodies but form only a minor part of the sequence. The boulder sheets dip northwest and are gently folded along northeast-southwest axes as in the terrain to the east. Clast imbrication indicates a predominantly southward flow for the powerful stream transporting this debris. The conglomerate forms the base of a fining-upward cycle with the overlying sands and interbedded violet mudstones that terminate in a flaggy limestone to marl before being overlain by more sands and interbedded pebbly sands and finally a second cobble to boulder conglomerate that is matrix-supported with interbedded sand. This second conglomerate contains only rounded river-transported clasts. At Baizhanggou village, about 2 km north of Yushe, the lower conglomerate seems to finger out into pebbly sands suggesting that the boulder conglomerate is a facies of the finer deposits to the east and thus part of the Mahui Formation. The basal conglomerates in the Liutan embayment are similar in lithology and could represent a tongue of the coarse sequence.

Interpretation. The oldest Tertiary deposits widely distributed in the Yuncu and Tancun subbasins constitute the Mahui Formation. Locally it unconformably overlies valley-wall colluvium and the deposits filling a valley that empties into the southwestern part of the Tancun subbasin. The full extent of the latter deposits was not explored in the course of this work beyond observations on their westernmost occurrence and contact relationships with the Gaozhuang Formation. No other deposits of this nature were observed in the area studied. The age of the Mahui Formation is Late Miocene, based on its vertebrate fossils (see Sect. 3.2).

The Mahui Formation was deposited, primarily as a fluviatile facies, in the valley of the trunk stream that is occupied today by the Zhuozhang River. This Mahui paleoriver is represented by a coarse facies, locally including boulder conglomerates, which fills a paleo-canyon and rapidly pinches out westward into festoon cross-bedded sands. The lower sandy part of the formation gives way laterally (and upward) to interbedded sands, muds and marls as the most distal facies. This array of facies, although not so coarse-grained, is approximated along the present-day Zhuo-zhan-ghe thalweg, so that the present river makes a reasonable model for the Miocene one. The only lacustrine facies, albeit a shallow-water one, is indicated by the widely traceable thin marls that frequently contain gastropod and ostracod remains, and at some sites, disarticulated fish bones. These are best developed at the top of the Mahui Formation type section in the easternmost Yuncu subbasin where concretionary calcareous claystones and flaggy limestones form an interval 9 m in thickness. The distribution of this facies seems distal to the fluviatile facies and may represent floodplain ponds rather than basin-wide lacustrine phases as postulated by other workers. We cannot demonstrate that the capping marl of the type section is represented by a similar facies throughout the Yuncu or Tancun subbasins largely because the erosional truncation of the Mahui seems to have removed the uppermost beds over a wide area. Outcrops near the mouth of the Donggou, west of Nanwang village, illustrate this truncation clearly.

We found little evidence of a westward shift of the major path of the Mahui trunk stream during deposition of that unit, which was suggested by Shi (1994). Boulder conglomerate in the lower part of the type section suggests a western position of the river axis early in deposition. The upward fining of the whole Mahui Formation was a response to gradually rising base level in the drainage system and hence a lowering of hydraulic flow.

3.5.2 Gaozhuang Formation

Description. Defined by Qiu et al. (1987), this unit occupies about half of the Yuncu subbasin and has a more limited, but significant, occurrence in the Tancun subbasin. The

Fig. 3.7 View (*eastward*) of Mahui-Gaozhuang angular unconformity exposed northwest of Beimahui. Dipping carbonate-cemented mudstones of the *upper part* of the Mahui Formation are overlain by more gently-dipping cemented sandstones of the basal Gaozhuang Formation

Yuncu subbasin type section extends from Taoyang nearly to Zhaozhuang village, a distance of about 6 km, and includes 400 m of strata composed of more than 70 % sand bodies and the remainder primarily mudstones with minor marls. Three fining upward cycles are recognized within the Gaozhuang Formation (Fig. 3.8). The lowest, the Taoyang Member, assembled from the CAL, CAT, CAS, CAGu subsections, makes up almost the entire formation. The upper part of the Gaozhuang Formation is broken into two units: the Nanzhuanggou Member, assembled from the CAH, CAJ, CAG and CANa subsections, overlain by sand and gravel of the basal part of the largely truncated Culiugou Member (CAG, CANa, CANas, CAC and CACs subsections). Less than 30 m of the latter unit remains beneath the disconformably overlying Mazegou Formation.

In the eastern Yuncu subbasin the Gaozhuang Formation lies with angular unconformity on the Mahui Formation (Fig. 3.7), but to the west, in the vicinity of Dongzhuang village, these units pass into apparent conformity. On the western side of the Donggou and westward to near Dongzhuang the basal beds of the Gaozhuang Formation are a series of up to three stacked stream channel fills, individually about 30 m thick, of cross-bedded yellow (10YR6/14) pebbly sands grading upward to thick-bedded yellow sands

with thin violet (5R/4) claystone interbeds. Up to 2 m of local relief occurs at the base of these channel fills. Pebbles and cobbles constitute the coarsest fraction and are predominantly clasts of Triassic sandstone and intraformational clayballs, but minor amounts of pebble-sized quartz and chert are also present. Further west these gravels grade into cobble to boulder conglomerate well exposed in the Hujiagou just east across the Gaozhuang valley from Taoyang village. However, north of Taoyang, in the CAT subsection, equivalent strata, correlated paleomagnetically, are fine to medium cross-laminated sands comparable in grain-size to those in the easternmost Gaozhuang outcrops in the Yuncu subbasin. We suspect that the lower Gaozhuang, and perhaps upper Mahui Formation is represented in the Taipingou reentrant south of Taoyang based on the presence of yellow pebbly sands there.

Most of the Taoyang Member of the Gaozhuang Formation is composed of fine to medium sands, usually showing low angle cross-lamination and occasionally marked by pebble to cobble lenses of lithic clasts (Triassic debris and autocycled material) interbedded with thin violet mudstones or claystone. The sequence is composed of many couplets of sand grading to mud. A notable local coarsening occurs nearly 300 m above the base in the southeastern Nanzhuanggou where cobble to boulder conglomerates again come into the section. The Honggou subsection (CAHo) contains the eastern end of this lens of conglomerate which thickens to 10 m to the west as shown by outcrops in the Yanjiagou and westward into the northern wall of the Dongwa just southeast of Baihai, but returns to sand in the outcrops around the latter village. The clasts range from rounded pebbles to boulders, dominantly of Triassic debris, but with locally abundant clay balls, and minor components, no larger than cobbles, of quartz, quartzite and black chert. These conglomerates grade upward into festoon cross-laminated, yellow fine to medium sand with lenses of violet, dark violet and light greenish-gray claystones.

About 20 m above the base of the conglomerate lenses in the Shagou (continuation of the CAHo subsection) a 2–4 m thick prominent green-gray marly claystone occurs, containing white calcareous nodules and lenses of flaggy white limestone. This marl can be correlated paleomagnetically and by a widely traceable marl in the vicinity of Gaozhuang village where it is recorded in the CAH, CAJ and CAG subsections. A second marl of similar character occurs about 15 m higher in the section in the CAJ and CAG subsections and can also be widely traced north and east of Gaozhuang village. Both of these marls may be locally removed by scour during emplacement of sand bodies above them but both can be mapped westward to the mouth of the Nanzhuanggou where they dip beneath the modern alluvium of the Zhaozhuanghe. (The small Zhuozhang stream drains from Zhaozhuang village southeastward to Baihai and Beicun).

These marls (Fig. 3.8) mark a shift to lacustrine conditions at the top of the Taoyang Member and the base of the Nanzhuanggou Member. Above them, cropping out widely on the floor of the Nanzhuanggou, is a 46 m thick body of massive-appearing gray to dark gray, yellow-mottled claystone rich in lignite and the remains of aquatic molluscs and ostracods. Flaggy and irregular carbonate concretions occur at several levels as do thin bodies of yellow fine sand. Comparison of the Nanzhuanggou subsection (CANa) with that northwest of Gaozhuang (CAG) shows that this lacustrine body fingers northward into fluviatile sands near the northern margin of the Yuncu subbasin. Likewise, traced laterally to the west to the Zhaozhuanghe and further into the Mazegou north of Baihai, these Nanzhuanggou Member dark clays become interbedded with thicker, and often cross-bedded sands and assume a more oxidized violet color like that of the mudstones commonly interbedded with sands in the fluviatile sediments of the Yushe Group.

A thick (10–20 m) unit of cross-bedded yellow fine sand with lenses of granules and pebbles of predominantly Triassic clasts and rarer quartz and chert and lenses or thin beds of violet mudstone mark the return of more extensive fluviatile environments in the central Yuncu subbasin and the initiation of the upper cycle of sedimentation referred to

as the Culiugou Member of the Gaozhuang Formation. This unit crops out on the western wall of the Nanzhuanggou and northwestward into the Culiugou east of Zhaozhuang. The basal sands are overlain by 10–15 m of interbedded thin yellow sands and violet mudstones before being disconformably truncated by the overlying Mazegou Formation. This thin remnant of the Culiugou Member laps onto faulted Triassic sandstones north of the Culiugou and extends southwestward, striking across the type section (CANas, CAC and CACs subsections) into the mouth of the Mazegou to crop out around the Baihai and Beicun villages.

The Gaozhuang Formation has been mapped in two places in the Tancun subbasin. More than 45 m of strata composing one fining-upward cycle rest with apparent conformity on the northwest-dipping Mahui Formation 0.65 km east of Jiayucun (the CAUl subsection). The basal 22 m are festoon cross-bedded sands with lenses of pebble to cobble conglomerate made up mostly of rounded Triassic clasts and minor quartz and chert. There are scattered subangular boulders of Triassic sandstone in the basal part of this conglomerate as well as large clayballs of violet claystone. The conglomerates are succeeded by 25 m of violet mudstone with thin medium through very-fine sand interbeds. Traced to the north these sand bodies become thicker cliff-forming cross-bedded granule and pebble-bearing sands (Fig. 3.9). A thin marl occurs in the highest outcrops of the Gaozhuang Formation in this area only a few meters above the top of the CAUl subsection. The Gaozhuang Formation traced northward from the CAUl subsection oversteps the Mahui Formation and locally rests on a basement ridge. Farther north the contact between these units passes under loess just southeast of Yushe town. To the south it forms the highest outcrops at the head of the Daxigou, 1 km south of Jiayucun (village), and passes westward beneath loess and the alluvium of the Zhuozhanghe (river).

Flat-lying pebbly sands lie just below the loess on the crest of the ridge separating the westward drainages of Liyücun (north) from Dengyucun (south). This ribbon of sand can be traced southeastward across the southward extending ridge of Triassic basement to a northeast-southwest oriented outcrop lapping the eastern side of the Triassic salient. The western end of this outcrop is about 10 m thick and composed of cross-bedded pebbly sands with calcite-cemented concretions whose base lies at about 1050 m elevation. One and seven tenths of a kilometer to the southeast, at about the same elevation, 12 m of flat-lying, cross-bedded, pebbly medium-grained, yellow sands are surmounted by more than 5 m of thin-bedded planar siltstones that grade to violet claystone before being truncated by loess. Further east these flat-lying deposits (Fig. 3.10) rest on the top of the Triassic salient at 1080 m where 40 m of cross-bedded gravels overlain by violet mudstones cross the ridge to its eastern slope to lie on the

Fig. 3.8 View to northwest of the escarpment above Gaozhuang village, showing the Nanzhanggou Member mudstones containing three interbedded marls within the Gaozhuang Formation (CAG section); trees provide scale

Fig. 3.9 Looking north-northwesterly at a nearly complete local section of the Gaozhuang Formation in the northern Tancun subbasin near Jiayucun. The basal Taoyang Member forms light-colored outcrops of pebbly sands at the base, behind the trees, overlain by thin-bedded claystones of the Nanzhuanggou Member

Fig. 3.10 Looking north in the southeastern Dengyucun Valley, southern Tancun subbasin, the angular unconformity between the horizontal Gaozhuang Formation and the west-dipping underlying Mahui Formation in the foreground and *middle left* above trees is apparent

boulder conglomerates of the basal Mahui and perhaps pre-Mahui sedimentary fill of the valley to the east. In these eastern outcrops the upper portion of the gravels contains pedogenetic carbonate nodules. The cumulative thickness of these deposits approximates 70 m. They are paleontologically correlated with the Gaozhuang Formation for they bear the same rodent fauna contained in the lower part of the Gaozhuang in the Yuncu subbasin and in the remnant found in the northern Tancun subbasin.

Interpretation. The Gaozhuang Formation is principally a fluviatile unit aggrading in the course of the Zhuozhang River in the Tancun subbasin and major tributaries from the Yuncu subbasin and the valley entering the Tancun Subbasin from the southeast. Prominent cobble to boulder conglomerate lenses at the bottom and near the top of the Taoyang Member in the Yuncu subbasin can be interpreted as axial facies of the ancient Yuncu River which maintained a southeasterly course just north of its present-day counterpart. The occurrence of shallow water and sometimes saline lacustrine deposits only at the top of the Taoyang Member and the larger proportion of mudstone and claystones there and in the overlying Yushe Group sediments suggests that a lowering of gradients and ponding of the axial drainage was initiated in the Yuncu subbasin towards the close of Gaozhuang deposition.

The Taoyang Member is not very fossiliferous, especially in its lower part, so faunal data are lacking to pinpoint its age. In contrast, the numerous vertebrate species from the younger Nanzhuanggou and Culiugou members indicate early Pliocene age for the upper half of the Gaozhuang Formation.

We found evidence of equivalents of the Gaozhuang strata in the northern and southeastern Tancun subbasin. To the north these represent a continuation of the interplay of sedimentary environments of the Mahui Formation with coarsening of the basal gravels in the Gaozhuang. We postulate that the upper sheet of thick boulder conglomerate north of Yushe town ("ucgl", Fig. 3.1) represents the base of Gaozhuang equivalents that overlie the Mahui Formation and Triassic basement in the ancient canyon linking the Tancun with the Ouniwa subbasins. In any event the continuation of the fluviatile regime indicated by the Mahui Formation clearly carries into the Gaozhuang, indicating the stability of axial drainage of the Zhuozhang River. Likewise to the south, the Gaozhuang drainage emerging from the valley east of Taiqiu passes northwest toward a union (now broken) with the Zhuozhang River. This tributary carried coarse debris and forms the highest Yushe Group outcrops, which cross a ridge of Triassic sandstone. No younger Yushe Group sediments

Fig. 3.11 Upper part of the Mazegou Formation at locality YS 99, view to east. Gray silts and clays dominate the section. Here a meter-thick sand body is offset ("down" to lower right) by a fault. The beds dip gently and are overlain by loess at the top of the hill. Photographed by L. Flynn, September 16 1988

have been recognized in the eastern Tancun subbasin, but events in the Yuncu subbasin, as interpreted above, indicate continued axial drainage at least until the close of Gaozhuang deposition. Most of the Gaozhuang sediments that existed in the Tancun subbasin must have been largely removed during the Quaternary except in the axial drainage boulder conglomerates north of Yushe town. The Gaozhuang Formation forms the major part of the fill of the Ouniwa and Nihe subbasins to the north.

3.5.3 *Mazegou Formation*

Description. This is the third major unit in the Yushe Group as defined by Qiu et al. (1987) in the Yuncu subbasin. The unit is confined to the western part of that subbasin. Its type section extends from the mouth of the Langzhanggou 0.8 km southwest of Zhaozhuang, northwestward to the Wanwanggou 0.5 km west of Liujiagou village, a distance of about 4 km (measured subsections CAZ, CAX, CAY, CANg, CACh and CALj). Although its describers were impressed with its coarser clastic nature and cliff-forming habit in comparison with the Gaozhuang Formation, our measured traverse through the middle of the Mazegou outcrop belt indicates that it contains thick mudstone units (about 55 % of total thickness) with interbedded poorly sorted lenticular sand bodies. The total thickness of this unit is about 170 m. It appears to be conformable on the Culiugou Member of the Gaozhaung Formation although our magnetostratigraphic analysis indicates that a significant hiatus (about 0.5 Myr) is

present at this contact. Lithologic contrasts across this contact are often subtle due to superposition of similar facies. The yellow Mazegou sandstone bodies are lenticular; only the thickest have any substantial lateral continuity. On average the grain-size of these sands is fine to medium, but coarser units were observed especially at the top of the unit (subsection CALj). They are usually cross-laminated and contain lenses of granule to pebble-sized clasts, mostly derived from the Triassic basement, but containing greater amounts of quartzite, black chert and red jasper than often seen in older Yushe Group units. More conspicuous is the frequent occurrence of autocycled claystones and limestones as a result of the cut and fill nature of the Mazegou sands. The claystones and mudstones are often darker in color (5R4/2-10R4/2) than those of the Gaozhuang Formation and range to dark gray and even black, and locally carry lignitic debris (Fig. 3.11). These darker clays often have small aquatic molluscs. Carbonate units are rare; only lenticular marls and nodular carbonate of limited lateral continuity are present. Carbonate cementation of sand bodies is also rare.

The Mazegou Formation can be traced northeastward into the head of the canyons north of Gaozhuang village where greenish-gray and violet claystones with thin, lenticular sands, and a locally mappable marl rest on the Culiugou Member of the Gaozhuang Formation and lap onto Triassic basement farther north. Like the Gaozhuang Formation, fine-grained units of the Mazegou Formation passively fill steep-sided canyons cut in the basement north of Shencun. In the southwestern part of its outcrop, pebbly sandstones and interbedded claystones of the Mazegou Formation lap onto basement rocks in the Baozigou just west of Yuncu town. The youngest part of the unit crops out beneath the extensive loess cover west of Liujiagou (subsection CALj) and at another site a kilometer to the north.

Interpretation. The Mazegou Formation accumulated under a lower energy regime than older Yushe Group units. The mud-dominated, thin-bedded and lenticular nature of its elements suggest a lateral-accreting system in contrast to the vertical accretion of fining-upward cycles typifying the Mahui and Gaozhuang formations. This pattern can be compared with observations on sections of the Old Red Sandstone made by Allen (1964). There are no coarse sands and conglomerates marking the axis of drainage of the Yuncu subbasin except perhaps in the uppermost part of the unit in its northwestern outcrop. These facts suggest that the Mazegou Formation may have been impounded in the central Yuncu subbasin, more like the overlying Haiyan Formation than the externally draining older Yushe Group units. Hence, the prevalence of dark clays and muds with molluscs and plant debris, indicates a shallow water lake or floodplain at or near water table. This lithology is approximated only locally in the Gaozhuang Formation at the top of the Nanzhuanggou Member. The fossiliferous Mazegou Formation yields vertebrates derived with respect to the Gaozhuang Formation and they demonstrate middle to late Pliocene age.

3.5.4 Haiyan Formation

Description. Qiu et al. (1987) described their Haiyan Formation from three traverses: (1) a long section from Liyucun in the Tancun subbasin through Haiyan in the western Yuncu subbasin; (2) a shorter section from Hewa westward through Haiyan to basement at Shiniushan; and (3) a short section in the northwestern Yuncu subbasin just north of Xizhoucun ("cun" = village), which afforded a possible contact with the Mazegou Formation. The evidence indicates that the unit was restricted to the western Yuncu subbasin and that it consisted of "greyish-yellow sands and violet or green clays" (Qiu et al. 1987), but doubt about the nature of the lowest strata and their contact resulted in only tentative assumptions of the unit thickness. We estimate the thickness at more than 60 m.

Most of the Haiyan Formation consists of siltstones and claystones (about 70 % of outcrop), ranging in color from tan through yellow to green and violet. These deposits contain aquatic molluscs and plant remains. Interbedded are lenticular bodies of yellow friable medium sand, often cross-laminated, and containing lenses of granule-pebble conglomerate (Triassic sandstone, quartzite, and black chert clasts). A typical view of the horizontal Haiyan Formation is presented in Fig. 3.12.

The present investigations have clarified the nature of the Haiyan Formation and we can now report that this unit is flat-lying, and remnants extend eastward to the divide between the Yuncu stream tributaries at Damalan and Zhaozhuang. There it rests with angular unconformity on successively older parts of the Mazegou Formation on a low relief surface that ranges from 1070 to 1080 m in elevation. These transgressive strata are thin-bedded yellow concretionary siltstones mentioned by Qiu et al. (1987) as possibly the basal strata of the Haiyan Formation. Fortunately this distinctive lithology is widely recognizable in the western Yuncu subbasin. Outcrops on the western side of the Yuncu subbasin from Zhengzhaigou (subsection CAZh) north to near Haiyan (Liupinggou subsection, CAP) can be correlated lithically and record at least 54 m of fluviatile and lacustrine strata with the characteristic yellow concretionary siltstone in the upper 6 m extending to about 1080 m elevation. Thus the marker bed siltstones occur in the upper part of the Haiyan Formation although locally forming the basal part of the unit to the east as Qiu et al. (1987) proposed. In the northwestern part of the basin between Xizhoucun and Qingyangping the CAQ subsection begins close to Triassic basement with the marker siltstone lithology (at about 1050 m elevation) at the base of a 40 m section overlain by fluviatile strata. The CAQ section extends the composite for the Haiyan Formation to about 75 m (refer to Fig. 3.4d).

Interpretation. The general lithology and sedimentary structures observed in the Haiyan area closely resemble those of the Mazegou Formation, except for the greater quantity of silt in the Haiyan Formation. The range of depositional environments is also similar to that of the Mazegou Formation. The Haiyan Formation is the uppermost Yushe Group unit in the Yuncu subbasin and has the most westerly positioned depocenter. The dominance of mud and lack of coarsely clastic channel fills within this unit indicate the relatively low gradients of surrounding basement terrain. A significant quantity of silt in this unit may be of aeolian origin as the Haiyan Formation is contemporaneous with deposition of the early Matuyama Wucheng Loess. The Haiyan *Equus* fauna demonstrates Early Pleistocene age correlation.

3.5.5 Yuncu Subbasin Composite Stratigraphy

The 28 lithological sections of the Yuncu-Tancun areas of Yushe Basin (Fig. 3.13) comprise about 800 m of Late Neogene strata. Our research approach was to sample the sections for magnetic analysis (Fig. 3.14) at the same time as they were measured and described, enabling unambiguous polarity assignments to individual strata, and ultimately

Fig. 3.12 Horizontal greenish clays and silts of the Haiyan Formation at locality YS110. Gen-Zhu Zhu sits atop the small hill above the locality. These beds are interpreted as mainly lacustrine in origin. In the background, Lishi and Malan loess striped with paleosols drapes over the paleotopography. Photo to north by L. Flynn, September 20 1988

clear age estimates for fossil occurrences. The 28 sections are aligned by both geological and magnetostratigraphic evidence to provide a robust, datable stratigraphy. The magnetic stratigraphy developed in the following chapter complements the general age assignments for Yushe formations, based on vertebrate fossils. The paleomagnetic data confirm that the Miocene–Pliocene boundary occurs high in the Taoyang Member of the Gaozhuang Formation.

3.6 Yushe Group in the Ouniwa and Nihe Subbasins

The Ouniwa and Nihe subbasins lie to the north of the Yuncu and Tancun study areas (Figs. 1.2, 2.3). While not highly fossiliferous, important specimens have been found from each subbasin. Basal deposits are younger than the oldest fill of the Yuncu subbasin, and the few faunal elements from Ouniwa and Nihe appear to derive from higher levels, estimated to be Pliocene in age.

3.6.1 Ouniwa Subbasin

Description. The Ouniwa subbasin is broadly connected southward with the Tancun subbasin through a 3 km wide ancient canyon cut across the Triassic ridge that lay between them (Fig. 3.1). This canyon was part of the pre-Yushe Group drainage as Mahui Formation sands and gravels extend through it. The Ouniwa subbasin fill lies mostly east of the present Zhuozhanghe and extends up tributaries cut into the Triassic bedrock to the north and east.

Teilhard de Chardin and Young (1933) were the first geologists to examine the fill of this subbasin and their concept of what we now call the Yushe Group was formed from a section that extended eastward from Houmu (now Gengxiu) village onto the flanks of basement inliers south of Ouniwa. This section samples only the upper part of the subbasin fill, but it remains a valuable record of the stratigraphy and biostratigraphy for this basin. Later, the Regional Geological Survey Brigade of the Geological

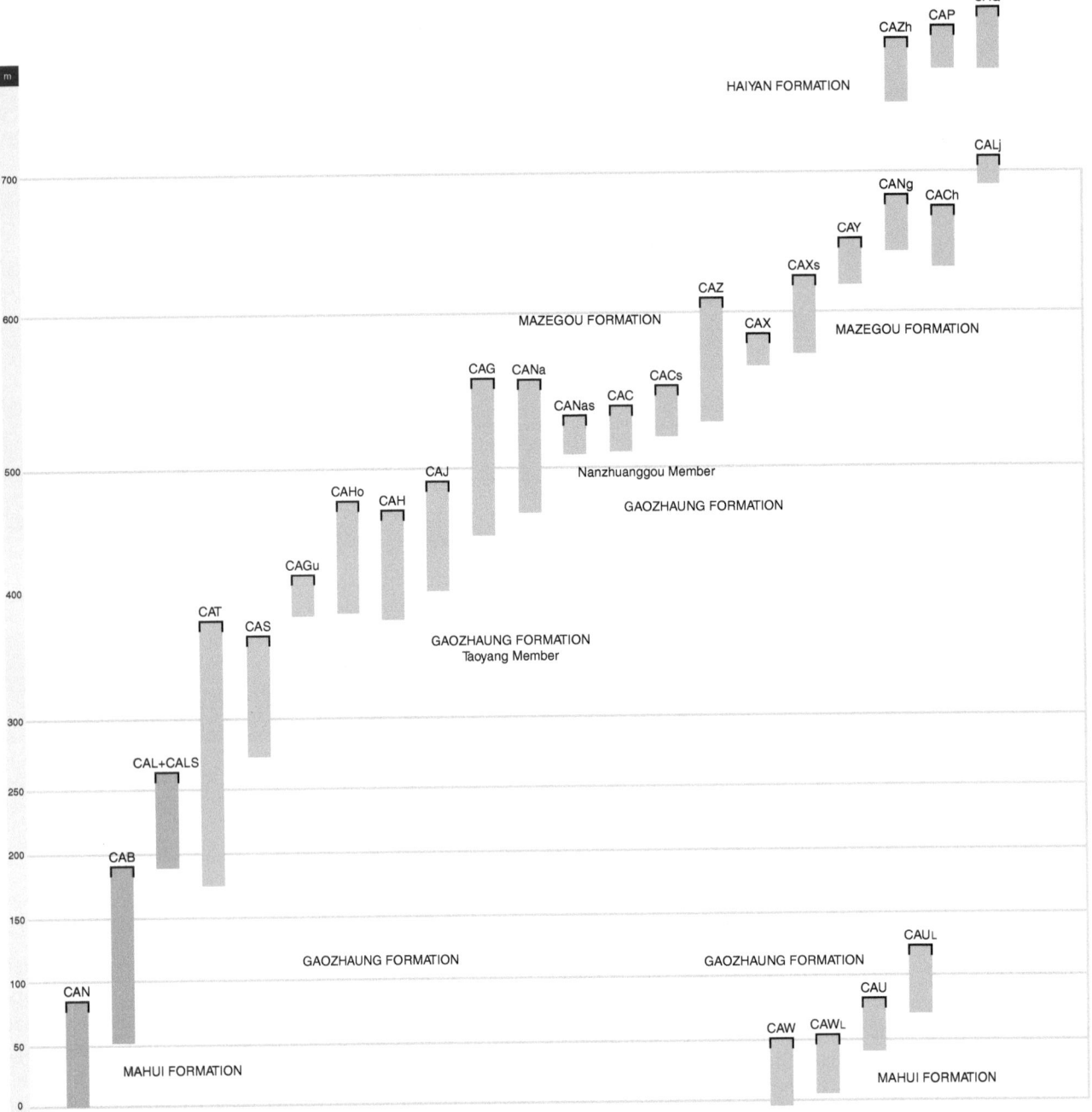

Fig. 3.13 All Yuncu Subbasin sections correlated and aligned on the 800 m composite stratigraphic section, with the four Tancun subbasin sections correlated in the *lower right* quadrant by means of the Mahui-Gaozhuang unconformity

Bureau of Shanxi Province produced a series of 1:200,000 geological maps (1976) and summarized the geology (1985) of the Yushe Basin. They measured a 340 m thick section in the Ouniwa subbasin, the lower 200 m of which were interbedded sands and gravels resting on the Triassic at Zhaowang village. The upper 140 m were interbedded sands and clays near Gengxiu and Ouniwa where similar lithologies were recorded in Teilhard de Chardin and Young's (1933) section which also lies above the conglomerates. The latter estimated that the whole section

probably exceeds 200 m. Our own observations are generally in accord with those of previous authors.

Reconstruction of the stratigraphy of the Ouniwa subbasin from the above sources and our investigations leads to the following composite sequence:

Unit 1: The lowest boulder conglomerate (lcgl, Fig. 3.1) rests on the Triassic on the flanks of the ancient canyon north of Yushe and can be mapped across the canyon where it rests on and interfingers with pebbly sands of the Mahui Formation of the northernmost Tancun subbasin. This unit fines upward to

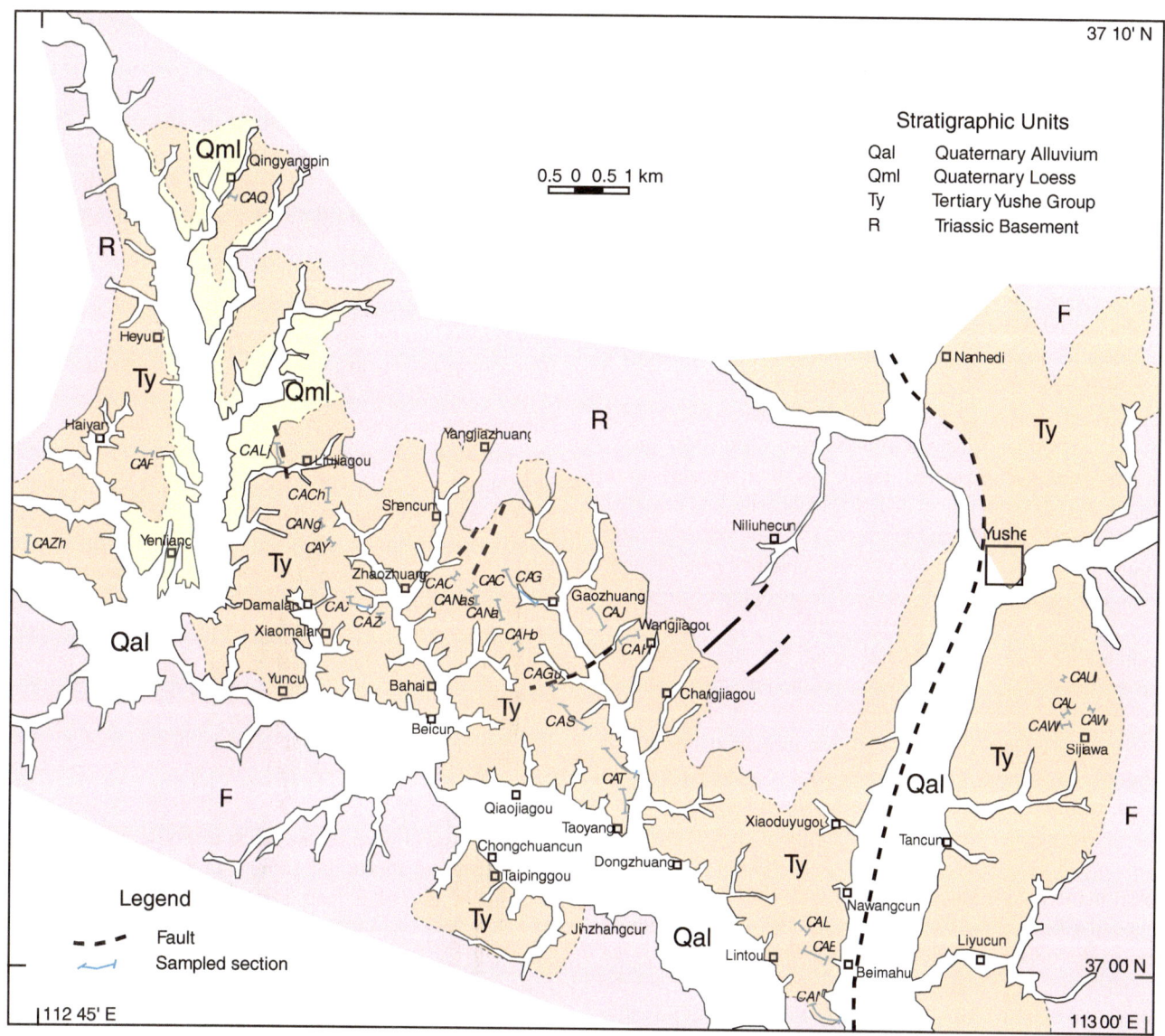

Fig. 3.14 Map of the Yuncu subbasin of the Yushe Basin, showing paleomagnetic sections, which are indicated by *blue irregular lines* with letter designations corresponding to measured sections of Fig. 3.13. Faults are indicated by *dashed black lines*. Figure from Chap. 4, thanks to Kainian Huang and Neil Opdyke

pebbly sands, sands, clays and finally to flaggy limestones and marls forming a cycle of about 100 m thickness. Because of its interfingering relationship we believe this unit should be included in the Mahui Formation.

Unit 2: As observed in deep southcutting canyons of the Xiaodougou, Hulugou and Maoshuigou on the eastern side of the canyon south of Qingyu village, unit 1 is overlain by sands with lenses of pebbles and cobbles of Triassic sandstone and conspicuous red claystone granules and pebbles that fine-upward to violet mudstone but lack a capping marl. This unit is also estimated to be about 100 m thick.

Unit 3: In the Maoshuigou units 1 and 2 are overlain by a second boulder conglomerate (ucgl, Fig. 3.1). This unit contains interbedded lenses of sand and lacks subangular large clasts

being mainly a river-transported conglomerate. Imbrication indicates transport to the south. Its top is truncated but is seen to grade upward to sand and violet mudstone as in underlying units. Traced into the Ouniwa subbasin to the north this unit appears to be the lower conglomerate body resting on the Triassic east of Tianjiagou. These conglomerates pass laterally westward and upward into thick-bedded sandstones, pebbly sandstones and thinner violet mudstones, the whole being about 100 m in thickness.

Unit 4: Northwest of Zhaowang village, a third boulder conglomerate crops out lapping onto basement of the eastern rim of the Ouniwa subbasin. This unit contains rounded river-worn cobbles and boulders as well as large angular corestones all derived from the Triassic countryrock. We believe this is the conglomerate at the base of Teilhard de Chardin and Young's (1933) section ("layer 1 of the Fig. 7"). Above this are gray and violet gravelly sands ("Layer 2"), dark violet clays ("Layer 3"), violet sands ("Layer 4") and yellow sands capped by 5 m of marl and chalky limestone ("Layer 5"). The latter forms the top of the upward-fining cycle of the youngest mappable boulder conglomerate in the Ouniwa Subbasin. This unit is at least 100 m in thickness.

Unit 5: Above the marl and limestone at the top of unit 4 are thick-bedded sands and pebbly sands with thinner bodies of violet mudstone. This is Teilhard de Chardin and Young's (1933) "Layer 6." It crops out in the northwestern part of the subbasin and in the reentrant in the basement around Ouniwa village. This unit is probably close to 50 m thick.

Interpretation. Scattered fossil mammal remains were obtained from the Ouniwa and Nihe subbasins and reported in Teilhard de Chardin and Young's (1933) pioneering work. Young (1935) described the material obtained near Houmu (now Gengxiu) and later workers have obtained material from units 3 through 5 of our synthesis, none of which include taxa restricted to Mahui strata. For this reason, and the evidence of stratigraphic superposition described above, we believe it is justifiable to attribute units 2 through 4 to the Gaozhuang Formation, bringing that formation northward from the contiguous Tancun subbasin. It is possible that the strata above the mappable marl (Unit 5) belong to a distinctive cycle like the upper (Culiugou) member of the Gaozhuang Formation in the Yuncu Subbasin. Of the few fossil mammals obtained around Ouniwa in Unit 5 only the hyaenid *Chasmaporthetes* sp. is restricted to rocks equivalent to the upper Gaozhuang or Mazegou formations in the Yuncu subbasin. Extending this reasoning, we consider the age of the Ouniwa fossils to be early Pliocene, but possibly the earliest sedimentation began there at the end of the Miocene. The total thickness of the Gaozhuang Formation in the Ouniwa subbasin is probably at least 350 m, consistent with the 340 m measured by the Geological Bureau of Shanxi Province and the 400 m for the type section in the Yuncu subbasin.

In the Ouniwa subbasin the Gaozhuang Formation is dominated by high energy clastic deposits shed from local basement and carried in by high gradient streams from sources east of the basin. The major boulder conglomerate bodies are arranged along the eastern basin margin and interfinger westward with finer grained clastics. This facies accounts for two-thirds of the total thickness of the unit. The sediments of the higher part (Unit 5) showed a change in average grain-size to sands, silts, clays and carbonate units implying lower energy processes including shallow water bodies in the closing phase of deposition in the Ouniwa subbasin.

3.6.2 *Gengxiu Formation*

Description. The youngest water-laid deposits in the Ouniwa subbasin were recognized by Huang and Guo (1991b) as the Gengxiu Formation for more than 30 m of horizontal, predominantly lacustrine deposits, lying with angular unconformity on the Gaozhuang Formation of the western Ouniwa subbasin. The type section of their Gengxiu Formation lies 500 m north of the village (the old Houmu), adjacent to the railway line, where it is disconformably overlain by the Lishi and Malan loess.

Interpretation. Huang and Guo (1991b) correlated this unit with the Haiyan Formation of the Yuncu subbasin (although not in those terms) on the basis that both contained remains of the rootless zokor *Myospalax*. However, it is difficult to reconcile Teilhard de Chardin and Young's (1933) meager references to the occurrence of rodents in the Houmu section with the newly defined Haiyan Formation so that an age on that basis is tenuous. On the other hand the structural relationship and lithic resemblance of these units do recommend their correlation, and therefore the Gengxiu Formation would be Early Pleistocene in age.

3.6.3 *Nihe Subbasin*

Vertebrate fossils were obtained from this subbasin by Liu Chang-Shan and Bai in the earliest days of exploration for fossil mammals (1922; part of the Andersson collection), but study of the basin fill was not advanced until 1976 when maps at 1:200,000 scale were prepared by the Regional Geological Survey Brigade of the Geological Bureau of Shanxi Province followed by a summary report (1985) by the same organization. In 1979 one of us (Qiu) measured a section in the Nihe Subbasin and this was used in an informal field guide in 1980. Later Huang and Guo (1991b) published a detailed composite section in two segments, the first beginning "on a hillside by the highway at the mouth of the Nihe River" (near Nihekou) and continues "along the Nihe River and terminates at Nihezhang." This is essentially the same traverse as adopted by Qiu in 1979.

Description. The Nihe subbasin lies slightly southwest of the Ouniwa subbasin and is confined to the western side of the Zhuozhang River where it is drained by the Nihe on the north and on the south by the stream joining the Zhuozhang River at Xiakou village. It lies west of the ancient canyon connecting the Tancun with Ouniwa subbasins and is mostly separated from the latter by Triassic basement as seen in a limited outcrop just north of the mouth of the Nihe at Nihekou village.

Basement outcrops nearly completely encircle the Nihe subbasin, leaving only a 3 km-wide opening to the southeast between Nihekou and Xiakou. This opens into the western side of the ancient canyon connecting the Tancun and Ouniwa subbasins, yet there seems to be no evidence of a trunk stream in the Nihe Subbasin during deposition of the Yushe Group there. In fact the dominance of lacustrine strata in the lower part of the basin fill suggests that the Nihe subbasin may have been nearly isolated from the Ouniwa subbasin. Regional westward tilting and uplift on the Zhanghe Fault would have added to the separation. The situation is similar to the isolation of the Zhangcun subbasin which shows a similar stratigraphy to that of the Nihe subbasin.

In general the Nihe subbasin fill consists of a basal conglomerate and sands, about 180 m thick, locally lying on Triassic bedrock and overlain by a succession of fining-upward cycles that begin with sand and end with violet claystones and often marl. The latter array of facies is about 350 m thick, for a total thickness of 530 m, slightly thicker than the aggregate in the Ouniwa subbasin. Qiu et al. (1987) estimated the thickness as 550 m. Using the lithostratigraphic data provided by Huang and Guo (1991b) we can divide the basin fill of the Nihe subbasin into the following six fining-upward cycles.

Unit 1: Resting on local basement outcrops near Nihekou are 30 m of a basal colluvium of poorly rounded or subangular cobble to boulder sized clasts of Triassic sandstone. This grades upward to stream-laid rounded cobble conglomerate interbedded with pebbly sands and overlain by interbedded tan pebble to cobble conglomerate and yellow sands with lenses of violet mudstone. These coarse clastics are 159 m thick and grade upward to gray, black and greenish clays with gastropods and ostracods capped by light gray laminated gastropod-bearing marl. This unit corresponds to units 1–21 of Huang and Guo (1991b); the total thickness is measured as 184 m.

Unit 2: Beginning with 6.5 m of rusty yellow fine sand, this unit grades upward through interbedded yellow sands, pebbly sands and violet mudstone to yellow green clays capped by laminated marl. These beds correspond to units 22–37 of Huang and Guo (1991b); total thickness, 57.6 m.

Unit 3: Yellow fine sands and interbedded violet claystones form the basal 11.3 m, which is supplanted upward by interbedded greenish to violet clays, thin sands and several marls. The clays and marls are rich in ostracods and gastropods. This unit corresponds with units 38–62 of Huang and Guo (1991b); total thickness is 66.8 m.

Unit 4: Yellow fine sands and interbedded violet claystones 24.4 m thick grade quickly to ostracod and gastropod bearing violet clays and greenish clays with laminated marl. This corresponds with units 63–72 of Huang and Guo (1991b): total thickness is 32 m.

Unit 5: Yellow fine sands interbedded with violet claystones form the basal 7.7 m of this cycle. Above this violet mudstone and yellow fine sands are interbedded with greenish, gray and black clays containing several marls that are rich in ostracods and gastropods. This and Unit 4 have a higher proportion of thin sand beds than lower units. It corresponds to units 73–98 of Huang and Guo (1991b) of total thickness, 95.6 m.

Unit 6: This unit is dominated by yellow to purplish sands. The basal 10 m unit of multistoried cross-bedded sands and violet clay lenses grades upward to thick (40 m) yellow fine sands and clays with a diverse ostracod and gastropod fauna accompanied by pelecypods followed by sandy gravels and greenish clay and marls capped by sands and violet sandy clays. This corresponds to units 99–109 of Huang and Guo (1991b); its total thickness is 95.6 m. Unit 6 is the youngest part of the Yushe Group in the Nihe subbasin. It laps out onto the Triassic west of Nihezhang, and is overlain with profound disconformity by the Lishi and Malan loess.

Interpretation. Following the regional stratigraphy of the Geological Bureau of Shanxi Province, Huang and Guo (1991b) recognized unit 1 as the Renjianao Formation. The remainder of the section was assigned to the Zhangcun Formation, broken into two members at the contact of our units 4–5. This is essentially the point where sandstones begin to dominate the section and they continue to do so to the top. Thus the Yushe Group in the Nihe subbasin can be seen as composed of two major fining-upward cycles: units 1–3, 308 m in thickness as the lower cycle and units 4–6, 223 m in thickness as the incomplete upper cycle.

Fossil mammal remains from the Ouniwa and Nihe subbasins provide a means of correlating the two sedimentary deposits with the successions in the Tancun and Yuncu subbasins. From its discovery in 1922, the Nihe subbasin has continued to yield important fossil material, but like the early work in the Yushe Basin the collections were not localized stratigraphically beyond the geographic locations of the villages from which they were gathered. In neither subbasin are elements distinctive of the Mahui Formation obtained, although the coarse basal deposits are largely devoid of remains. More recent collecting has established the presence of two chronologically important myospalacine rodents in both subbasins. Huang and Guo (1991b) record the presence of *Chardina truncatus* in beds 31–32 of their measured section in the Nihe subbasin (Unit 2 of this work), and *Pliosiphneus* n. sp. (see Volume 2) was obtained by our party south of Tianjiangou (locality YS 36) in beds stratigraphically equivalent to unit 3 of the Ouniwa subbasin. Both rodents occur low in the Gaozhuang Formation of the Yuncu subbasin. Some taxa restricted to the Mazegou and Haiyan formations in the Yuncu subbasin also occur in historic collections from the Nihe subbasin including species of *Canis*, *Lynx*, and *Pachycrocuta*, but most of the fauna represent taxa that have ranges that include the Gaozhuang Formation as well.

Thus the temporal relationship of the fill of the Ouniwa and Nihe subbasins seems comparable to that seen in the Tancun and Yuncu subbasins corresponding with the westward shift in depocenter, i.e., Gaozhuang on the east (Ouniwa subbasin) and Gaozhuang to Mazegou to the west. For the Nihe subbasin, in conclusion, we apply the lithostratigraphic term Gaozhuang Formation to units 1–3, the lower cycle, and correlate units 4–6, the upper cycle of the Nihe Subbasin, with the Mazegou Formation. In temporal terms, the Nihe subbasin can be characterized as including deposits of early to late Pliocene age, including later stages of deposition than the Ouniwa subbasin.

3.7 Yushe Group in the Zhangcun Subbasin and Wuxiang Basin

The Zhangcun subbasin lies to the south of Yuncu Subbasin (Figs. 1.2, 2.3) and north of the Wuxiang Basin. Both the Zhangcun and Wuxiang regions are very important for their fossil content and figure into the history of paleontological exploration of Shanxi Province. Lithologically, the Zhangcun subbasin differs from the Yuncu Subbasin and its own stratigraphic nomenclature applies there. Zhangcun lake deposits are famous for their vertebrate faunas that represent moist habitat. Wuxiang Basin, downstream from Yushe Basin, contains Late Miocene valley fill with an important Baodean fauna.

3.7.1 Zhangcun Subbasin

Description. The Cenozoic section exposed in the Zhangcun subbasin was selected for detailed stratigraphic study by the Geological Bureau of Shanxi Province in the 1970s. This subbasin lies across the Wuxiang-Yushe county boundary and appears isolated from the Yuncu subbasin to the north, but may be connected with the northwestern Wuxiang Basin to the south.

The Bureau's stratigraphic study led to the conclusion that three formation-rank units could be recognized (in ascending order): Rejianao, Zhangcun and Louzeyu formations. This sequence consisted of three facies: alluvium, colluvium and coarse fluviatile deposits; finer fluviatile and lacustrine deposits; and the youngest unit represented by calcareously cemented clays and fine sands. This sequence was postulated as a sedimentary cycle filling a previously excavated topography with alluvium and coarser fluviatile deposits, the

impounded basin containing finer alluvium and lacustrine deposits followed by a shallowing lake with chemical sediments ending in the return to fluviatile environments. These conclusions were generalized into a basin-wide chronostratigraphy and the corresponding lithostratigraphic nomenclature was widely applied.

Although Huang and Guo (1991) had advocated subsuming the Louzeyu sediments into the Zhangcun Formation, they admitted that the "type locality is in the vicinity of the village of Louzeyu, Wuxiang" (Huang and Guo 1991: 23) as the lithostratigrahic name implies. However, they discounted previous contentions that those rocks lie with angular unconformity on the Zhangcun Formation. In a chapter in the same volume edited by Huang and Guo (1991), Rui-jin Wu (1991: 176–185) alludes to inclusion of the Louzeyu Formation as the "upper Zhangcun Formation", but still clearly shows its unconformable overlap on older strata in her cross-section (Fig. VII-4). We conclude that the Louzeyu Formation is a distinct lithostratigraphic unit tectonically separable from the underlying strata. Its designated type locality represents that unit, but the "type cross-section" mostly includes the underlying Zhangcun Formation through an error in correlation.

Interpretation. In 1994 Shi Ning published an extensive review of the Zhangcun subbasin lithostratigraphy, magnetostratigraphy, [10]Be flux chronology, palynology and environmental conclusions to present a more complete analysis of the subbasin's evidence for historical geology. Significant conclusions included a revision of the lithostratigraphy to break the Zhangcun Formation into two units: a lower Wangning Formation (new name) limiting the Zhangcun Formation to the top of its former span. Magnetostratigraphic study of the contacts at the top of the Renjianao and Zhangcun formations failed to show that these hiatuses represent significant time gaps. Over the years considerable confusion had been generated regarding the type locality and hence the stratigraphy of the Louzeyu Formation. Shi (1994: 26) clarified this and restored the formation to its uppermost position in the Zhangcun subbasin sequence.

The complete magnetochronology (Shi 1994; reviewed here in Chap. 4) indicated that the base of the Yushe Group in the Zhangcun subbasin is about 5.5 Ma just below the Miocene–Pliocene boundary; the Renjianao–Wangning boundary about 4.5 Ma; the Wangning–Zhangcun boundary between 3.4 and 3.5 Ma; and the Zhangcun–Louzeyu contact about 2.3 Ma near the Pliocene–Pleistocene boundary. The loess sequence lying disconformably on the Louzeyu Formation includes parts of the Wucheng, Lishi and Malan Loess. Wucheng Loess deposition ended about 1.5 Ma, the Lishi Loess deposition spanned approximately 1.0–0.6 Ma, and the Malan Loess encompasses 0.02–0.01 Ma at the end of the Pleistocene.

Shi (1994) reconstructed a pollen profile for the Yushe Group type section in the Zhangcun subbasin which yielded 10 pollen zones: Y1 upper Renjianao, Y2–Y5 Wangning; Y6–Y8 Zhangcun; and Y9–Y10 for the Louzeyu Formation. This profile indicated a climatic sequence beginning with moist and relatively warm climate in the earliest Pliocene, changing to a drying and cooling phase later in the Pliocene with the loss of subtropical components. By 2.3 Ma forest gave way to steppe and cooler savanna conditions ushering in Pleistocene environments coeval with major climatic changes elsewhere in the northern hemisphere.

The palynology in a more densely sampled section of Zhangcun subbasin was developed by a Nanjing team (Liu et al. 2002). A section through the clays and thinly bedded oil shales of the Zhangcun Formation offered the opportunity to observe continuous floral change for an interval of the Late Pliocene into Early Pleistocene, about 3.2–2.0 Ma on the current time scale. The dominant flora (Chenopodiaceae, Pinaceae, *Artemesia*, and *Ulmus*) is accented by less common thermophilic taxa (*Carya, Carpinus, Juglans/Pterocarya*), which together show a generally cooling trend through this interval (Liu et al. 2002). These authors note cyclical peaks in the thermophilic taxa, indicating possibly astronomical oscillations. The interpretation is consistent with the findings of Shi (1994), but the added detailed observations would not be apparent without the extraordinarily complete local record of deposition.

3.7.2 Wuxiang Basin

Description. Relatively little geological study of the fossiliferous Yushe Group sediments of the Wuxiang Basin has been conducted at a scale useful to this work. Nevertheless localities within a 7 km radius south and southwest of the county seat yielded material to the earliest collectors of the Geological Survey of China, much of which was described by members of the Sino-Swedish group. Teilhard de Chardin and Young (1933) traversed this region during their epic reconnaissance of the basin in 1932. They figured two representative sections and briefly discussed these (1933: 223–224, their Figs. 9, 10) from observations made on the road about halfway between Wuxiang and Qin towns. The localities are near Suhotze (*Sihezi*) village. They concluded that these deposits belonged to the same basin as those examined around Yushe. At Wuxiang, the basal deposits are thick boulder conglomerates resting on the Triassic while north and westward into the basin these are overlain by a succession of sands with interbedded violet clays and limestone units. Coarser clastics dominate this sequence rather like those they had observed near Houmu in the northernmost part of the basin.

Subsequent 1:200,000 mapping by the Geological Bureau of Shanxi Province (1976, Fig. 1.2) delineated the complex array of Triassic basement ridges and Yushe Group sediments contained in the intricate dendritic valleys of the pre-Yushe Group drainage. The Wuxiang Basin lies in the confluence of the Zhuozhang River draining south from the Tancun subbasin and the main tributary from the northwest. From that confluence the Zhuozhang River enters a deep canyon cut in the Triassic, and flows eastward and southward across the Taihang Shan to the North China Plain.

Interpretation. Deposits in the Wuxiang Basin must be several hundred meters in thickness including a thick basal conglomerate. As in the north, the rocks dip at low angles to the northwest, steeper near Triassic basement, and shallower as one ascends the section. Fossil collections were acquired from a dozen or more villages in the early work (intermittently from 1922 to 1931) by the Geological Survey of China. Judging from their geographic positions relative to the eastern basement margin, these sites must represent a significant stratigraphic span, yet they bear a rather uniform fauna of late Miocene age. Characteristic late Miocene, Baodean Land Mammal Age elements recovered from the Wuxiang Basin include the horses *Hipparion platyodus* and *Hipparion ptychodus,* the pig *Chleuastochoerus stehlini,* a giraffid *Honanotherium schlosseri,* and the deer *Eostyloceros blainvillei* and *Cervocerus novorossiae.* Coupled with absence of derived forms, productive Wuxiang Basin localities appear to be entirely Late Miocene in age.

Most of these sites are in outcrops referred to the Renjianao Formation by the Shanxi geologists. Confluence with the Tancun subbasin is nearly attained along the Zhuozhanghe so that the basal sediments of the Wuxiang Basin could as reasonably be referred to the Miocene Mahui Formation. The Wuxiang fauna has many similarities to that from the lower Mahui Formation, but subtle differences (e.g., presence of *E. blainvillei,* well known from Baode) suggest a still greater Late Miocene age for the rocks at Wuxiang.

3.8 Post-Yushe Group Deposition

Overview and Interpretation. By the close of Pliocene time the northern Yushe Basin was nearly filled with sediment, and the horizontal Louzeyu, Haiyan and Gengxiu formations accumulated locally. The occurrence of sediments of similar age to the south in the Qin Basin (Teilhard de Chardin and Young 1933) suggests that the same conditions must have been widespread in the Yushe Basin and elsewhere. Interfluves were widely capped by the oldest loess (Red loam of Teilhard de Chardin and Young 1933) that accumulated contemporaneously with the youngest Yushe Group water-laid deposits.

All of these younger rocks are still flat-lying, indicating a quiescence of tectonism since that time.

Rejuvenation of the pre-Yushe Group drainage began in the early Pleistocene, probably by capture of the Zhuozhang River course by streams breaching the tectonic barrier that previously impounded the drainage of the Yushe Basin. During this rejuvenation the Zhuozhang and its tributaries were widely superposed across basement ridges as they slightly departed from the precise course of their pre-Yushe Group predecessors.

Two major events altered the course of this rejuvenation of drainage. The first was marked by the thick accumulation of loess, filling newly excavated canyons. Dissection reestablished the drainage leading to near removal of loess canyon fills. This second rejuvenation was followed by further deposition of loess that blanketed the basin before finally being largely removed by the rejuvenation still underway at the present time.

In the following remarks we discuss each loess sheet including the first that accumulated contemporaneously with the youngest fluvio-lacustrine beds of the Yushe Basin.

3.8.1 Wucheng Loess

Description. The identification of this widespread unit in the Yushe Basin was first positively made by Shi (1994) from magnetostrataigraphic studies in the Zhangcun subbasin. However, Teilhard de Chardin and Young (1933), in their study of the section near Zhangjiagou (Zhangcungou) village on the western side of the Zhuozhang River, 10 km northeast of Wuxiang, had discovered a horizontally bedded dark Red "loam" (capital R) with a basal conglomerate unconformably overlying the westerly dipping Yushe Group (probably Mahui Formation). This was in turn disconformably overlain by a lighter colored red loam, its red paleosols dipping with the surface of disconformity. These deposits were disconformably overlain by "Loess." The latter two units were widely observed in the region.

Since that reconnaissance study, the regional mapping of the Yushe Basin by the Geological Bureau of Shanxi Province (1976) has carried the stratigraphy across the basin and identified the younger "red loam" as the Lishi Loess and equated the "Loess" with the Malan Loess. The older Red loam was referred to the Daqiang Formation, typified in the adjacent Zhongcun Basin, Tunliu County, farther south in Shanxi Province. Subsequent work by Di et al. (1984) and Shi (1994) demonstrated that the Red loam at

Wangning in the Zhangcun subbasin and elsewhere in the basin has the textural and structural properties of loess.

Loess deposition was punctuated by development of 8–10 deep red paleosols with iron-manganese films and small concretions but only a small amount of calcareous concretions. It was Early Pleistocene lower Matuyama in age, ranging from 2.5 to 1.5 Ma consistent with the few fossils found in the unit (e.g., at Zhangjiagou, Teilhard de Chardin and Young 1933). Shi (1994) correlated the Red loam with the Wucheng Loess of the Loess Plateau of Shaanxi Province, a conclusion we accept. The Wucheng Loess may be 30 m thick in the Yushe Basin. It was deposited on interfluves contemporaneously with the youngest basin fill, and unconformably overlies the older Yushe Group and Triassic basement. In the Yuncu subbasin, the older red loess correlated with Wucheng Loess fills the paleotopographic lows cut into the Yushe Group (Fig. 3.15).

3.8.2 Lishi Loess

Description. This unit, widely recognized on the Loess Plateau of Shanxi Province, blanketed the Yushe Basin following the earliest phase of rejuvenation of external drainage (background of Figs. 3.3, 3.4, 3.5, 3.6, 3.7, 3.8, 3.9, 3.10, 3.11, 3.12). Depending on topographic setting, the Lishi Loess may reach 50 m in thickness where it fills deep canyons. It is a brownish-yellow loess with 3–5 bands of brownish-red paleosols rich in calcareous concretions, the paleosol bands indicating the accumulation surfaces that often dip steeply in the fill of canyons. It can be seen overlapping the Wucheng Loess on interfluves, indicating that it filled the topography exhumed by the first drainage rejuvenation.

This unit was widely seen by Teilhard de Chardin and Young (1933) in their traverse of 1932. They designated it the "younger red loam" (indicated by lower case r in their figures) and assigned it a "Choukoutien" age. The Geological Bureau of Shanxi Province (1985) confidently assigned the Yushe Basin occurrences to the Lishi Loess defined in western Shanxi Province and correlated them with the upper part of the unit expressed in the Loess Plateau. Shi (1994), based on his paleomagnetic study of outcrops of this loess in the Zhangcun subbasin, showed that it was latest Matuyama (post Jaramillo) and Brunhes in age, 0.8–0.9 Ma to approximately 0.1 Ma.

Since accumulation of the Wucheng Loess in the Zhangcun subbasin (Shi 1994) ended at about 1.5 Ma, the hiatus between these loess sequences is about 0.6 Myr. This

Fig. 3.15 Outcrop of loess filling a paleo-gully carved into Yushe Group deposits. The deeply colored red loess (Wucheng Loess) fills an old channel; dark scars in shadow at the midlevel of the outcrop show where sediment was sampled by L. Flynn for screening for microfauna (locality YS83). Gen-Zhu Zhu stands on trail below. Photo by L. Flynn, September 5, 1988

represents the span of the first rejuvenation of topography in the northern part of the Yushe Basin.

the drainage of the Yushe Basin was again reexcavated and is continuing to evolve in a degradational regime.

3.8.3 Malan Loess

Description. This is the youngest and most ubiquitous loess sheet. It is a massive, yellowish loess, often with a basal gravel when deposited on stream valley terraces, and occasionally contains small calcareous concretions. It shows little structure except for vertical cleavage planes that break outcrops into columnar bodies. The Malan Loess blankets all topography and preexisting stratigraphy. It is 5–20 m thick, filling the deep canyons cut during the post-Lishi rejuvenation of topography in the Yushe Basin. Since the Malan Loess represents the glacial portion of the last glacial cycle, about 0.02–0.01 Ma, the hiatus occupied by the second rejuvenation of topography, the post Lishi-pre Malan span, would be about 0.08 Ma. During the Holocene

3.9 Structure and Cenozoic Geological History

3.9.1 Structural Features

Description of the structural features of the Yushe Basin must be restricted to those northern subbasins that were mapped during the course of our work. Teilhard de Chardin and Young (1933) provide some general conclusions about the late Cenozoic tectonics of a larger area in southeastern Shanxi Province, including the Yushe Basin, and these are in accord with our observations from the smaller area. Structural features, mostly in the Triassic, are shown on the Pingyao Quadrangle (1976) and these are indicated in Fig. 2.2.

As mentioned above, all but the final cycle of fluvio-lacustrine deposits in the Yushe Basin have a pervasive northwestward dip. Dip angles decrease upwards from about 15° for the Mahui Formation and lowest units of the Gaozhuang Formation, to 5–10° for upper levels of the Gaozhuang Formation and lower Mazegou Formation, to near horizontal for the upper part of the Mazegou Formation. All subbasins show an off-lap of stratigraphy in the dip direction. Geological maps covering the basin (Pingyao and Qinyuan quadrangles 1:200,000, Geological Bureau of Shanxi Province 1976) show that the basal Yushe units follow the strike of the northeast limb of the Qinshui synclinorium of Triassic age. This continued to be an axis of uplift during deposition of the Yushe Group.

Where mapped in more detail (1:50,000) in the northern Yushe Basin (Fig. 3.1), the location of the Mahui-Gaozhuang contact across the Zhuozhang River indicates a displacement of about 100 m, down to the east. For that reason we initially postulated a normal fault, the Zhanghe Fault, in the river course. An exposure of this fault was later found northwest of Yushe town, east of the narrow canyon in Triassic rock followed by the Zhuozhang River today. The outcrop shows Triassic on the southwest faulted against Mahui Formation boulder conglomerates marked by a 30 m wide zone of crushed Triassic sandstone trending N60°W. Fault-parallel drag folds were formed in the adjacent Yushe Group strata. This fault seems to define the western wall of the Zhuozhang River of Mahui-Gaozhuang time suggesting the structure has exerted syntectonic control of sedimentation in the northern subbasins.

Small faults of limited throw (up to 10 m) and short length were more commonly encountered cutting the older Yushe Group of the Tancun and eastern Yuncu Subbasins. Most show fault plane dips of 50–60°, and down to the east orientations like the Zhanghe Fault. Fault plane trends generally follow northeast axes and they break all units except the flat-lying youngest formations of the Yushe Group. Some can be traced from fractures in the Triassic interfluves across the valley-confined Yushe Group (e.g. northeast of Taoyang or west of Wangjiagou) suggesting they represent persistent movement of the gently folded Triassic basement. North of the Yushe Basin in the northwestern part of the Pingyao Quadrangle fault trends are more easterly paralleling the trends in the boundary faults of the Shanxi Graben system to the west. Offsets of Jurassic strata imply a component of strike-slip motion as inferred within the graben (Zhang and Zheng 1995).

Gentle folds of northerly trend are common in the Tancun subbasin and easternmost Yuncu subbasin confined to rocks of the Mahui and lower Gaozhuang formations. Most of these folds are subparallel or lie at low angles to the Zhanghe Fault, suggesting the influence of a strike-slip component in the movement of the large fault, as widely observed associated with the bounding faults of the adjacent

Shanxi Graben and North China Plain rift zones (Zhang and Zheng 1995). No faults were observed cutting the flat-lying youngest units in the northern Yushe Basin.

3.9.2 Cenozoic Geological History

Here we synthesize the Neogene geological history of Shanxi Province, specifically with respect to Yushe Basin. By late Miocene time the headwater tributaries of the Zhuozhang River were established in the axial region of the Qinshui Synclinorium and had cut deep canyons into the Triassic bedrock. At this time the steep gradients of these mountain rivers carried most of the debris eastward to the North China Plain. External drainage was reversed by an event or events around 7 Ma, most likely tectonic uplift in the Taihang Range along a northeast axis. This age estimate, rounded to the nearest m.y., is a minimum based on the oldest age documented for the Yushe Group. With the external drainage blocked, sedimentation began in the lowest lying part of the river system in the late Miocene. This event initiated sedimentation in the dendritic river valleys (subbasins) that are collectively grouped as the Yushe Basin.

In the northern subbasins the axis of the trunk stream is parallel to the axis of uplift so that this part of the system filled sequentially up gradient from south to north. The Tancun and eastern Yuncu subbasins have the oldest (Mahui) sediments while the terminal branches, the Ouniwa and Nihe subbasin records begin with the latest Miocene basal Gaozhuang Formation. At the same time continued uplift of the Taihang limb of the Qinshui Synclinorium translated the depocenters of Pliocene deposits northwestward into the heads of the western tributaries (Zhangcun, Yuncu and Nihe subbasins). Ultimately the subbasins filled with locally derived sediments. In the Zhangcun subbasin, Late Pliocene clays and oil shales accumulated in an uninterrupted sequence, offering fine scale detail of floral changes and possibly a record of astronomically controlled climatic oscillations (Liu et al. 2002).

Tectonic influence ceased at about 2.5 Ma and fine-grained flat-lying sediments accumulated in the final, northwestern-most depocenters. By this time the Triassic interfluves began to receive a blanket of loess (Wucheng) which extended across the older Yushe Group on the subbasin margins. Studies by Shi (1994) in the Zhangcun subbasin, show that the Wucheng Loess was deposited contemporaneously with the Louzeyu Formation, both units ceasing deposition at about 1.5 Ma, shortly after the Olduvai Chron.

At this point external drainage of the Yushe Basin was reestablished and the region shifted from an aggradational mode to a degradational mode as the gradient necessary for sediment removal was restored. Presumably slowing or cessation of uplift in the Taihang Shan allowed headwater erosion to breach the mountainous barrier to the east. Drainage patterns in the Yushe Basin had not been completely lost during Yushe Group sedimentation and streams began to reexcavate their courses with the establishment of gradients comparable to pre-Yushe time. The many examples of superposition of the Pleistocene rivers across bedrock ridges in the Yuncu and Ouniwa subbasins, for example, suggest that a southerly tilt of part of the terrain was established during this phase.

At about 0.9 Ma degradation slowed allowing deposition of sand and gravel in the Yushe drainage and eventually the accumulation of the Lishi loess on the interfluves and within the tributary watercourses. The Yushe Basin was again blanketed but this time mostly by aeolian processes. Only the upper part of the Lishi Loess is preserved in the Yushe Basin, perhaps representing only the last 0.5 Myr of its history (Shi 1994). At the close of Lishi time, about 0.14 Ma, degradational processes again dominated, removing much of the Lishi Loess from the drainage ways.

This degradational episode, like that of the early Holocene, bracketed the wide deposition of the Malan Loess in the Yushe Basin. These events are most likely to have been driven by climate change during the last glacial cycle. The last interglacial, like the present one, corresponded to a time of increased rainfall and removal of loess deposited during the preceding glacial. During the last glacial period the Yushe terrain was blanketed again most recently by the Malan Loess. Since then, Yushe Basin has continued its history of degradation.

References

Allen, J. R. L. (1964). Studies in fluviatile sedimentation: Six cyclothems from the Lower Old Red Sandstone, Anglo-Welsh Basin. *Sedimentology, 3*, 163–198.

Cao, Z.-Y., Xing, L., & Yu, Q. (1985). The age and boundary of magnetic strata of the Yushe Formation. *Bulletin of the Institute of Geomechanics CAGS, 6*, 143–153.

Di, L.-Xi., Cao, J., & Wu, R.-J. (1984). The genesis and age of the "R" Loess in southeastern Shanxi Province. *Acta Scientiarum Naturalium, Universitatis Pekinensis 5*, 88–95.

Geological Bureau of Shanxi Province. (1976). *Geologic Fenyang-Pingyao Quadrangle*. Beijing: Geological Publishing House.

Geological Bureau of Shanxi Province. (1985). *Late Cenozoic stratigraphy of Shanxi Province*. Beijing: Geological Publishing House (in Chinese).

Huang, B.-Y., & Guo, S.-Y. (1991). Stratigraphy. In B.-Y. Huang, S.-Y. Guo et al, (Eds.), *Late Cenozoic stratigraphy and paleontology from Central-Southern region of Shanxi* (pp. 1–70). Beijing: Science Press.

Licent, E., & Trassaert, M. (1935). The Pliocene lacustrine series in central Shansi. *Bulletin of the Geological Society of China, 14*(2), 211–219.

Liu, G., Leopold, E. B., Liu, Y., Wang, W., Yu, Z., & Tong, G. (2002). Palynological record of Pliocene climate events in North China. *Review of Paleobotany and Palynology, 119*, 335–340.

Lucas, S. G. (2001). *Chinese fossil vertebrates*. New York: Columbia University Press.

Pei, W.-Z., Zhou, M.-Z., & Zheng, J.-J. (1963). *Cenozoic Erathem of China*. Beijing: Science Press (in Chinese).

Qiu, Z.-X., Huang, W.-L., & Guo, Z.-H. (1987). The Chinese hipparionine fossils. *Palaeontologia Sinica, New Series, C25*, 1–243 (in Chinese with English summary).

Shi, N. (1994). The Late Cenozoic stratigraphy, chronology, palynology and environmental development in the Yushe Basin, North China. *Striae, 36*, 1–90.

Teilhard de Chardin, P. (1942). New rodents of the Pliocene and Lower Pleistocene of North China. *Publications de l'Institut de Géobiologie, Pékin, 9*, 1–101.

Teilhard de Chardin, P., & Leroy, P. (1945a). Les Félidés de Chine. *Publications de l'Institut de Géobiologie, Pékin, 11*, 1–58.

Teilhard de Chardin, P., & Leroy, P. (1945b). Les Mustélidés de Chine. *Publications de l'Institut de Géobiologie, Pékin, 12*, 1–56.

Teilhard de Chardin, P., & Trassaert, M. (1937a). The Proboscideans of South-Eastern Shansi. *Palaeontologia Sinica C, 13*(1), 1–85.

Teilhard de Chardin, P., & Trassaert, M. (1937b). Pliocene Camelidae, Giraffidae and Cervidae of South-Eastern Shansi. *Palaeontologia Sinica New Series C, 1*, 1–69.

Teilhard de Chardin, P., & Young, C. C. (1933). The late Cenozoic formations of S. E. Shansi. *Bulletin of Geological Society of China, 12*, 207–248.

Tucker, M. (1982). *The field description of sedimentary rocks. Geological Society of London Handbook*. Chichester: Wiley.

Wu, R.-J. (1991). Discussion on the sedimentary environment of the Yushe Group. In B.-Y. Huang, S.-Y. Guo et al (Eds.), *Late Cenozoic stratigraphy and paleontology from Central-Southern region of Shanxi* (pp. 176–190). Beijing: Science Press.

Young, C. C. (=Yang, Z.-J.) (1935). Miscellaneous mammalian fossils from Shansi and Honan. *Palaeontologia Sinica C, 9*(2), 1–42.

Zdansky, O. (1925a). Quartäre Carnivoren aus Nord-China. *Palaeontologia Sinica C, 2*(2), 1–25.

Zdansky, O. (1925b). Fossile Hirsche Chinas. *Palaeontologia Sinica C, 2*(3), 1–94.

Zdansky, O. (1927a). Weitere Bemerkungen über fossile Carnivoren aus China. *Palaeontologia Sinica C, 4*(4), 1–30.

Zdansky, O. (1927b). Weitere Bemerkungen über fossile Cerviden aus China. *Palaeontologia Sinica C, 5*(1), 1–21.

Zhang, D.-N., & Zheng, R.-S. (1995). Numerical simulation of the detachment dynamics in North China Basin. *Acta Seismologica Sinica, 8*(4), 511–517.

Chapter 4
The Paleomagnetism and Magnetic Stratigraphy of the Late Cenozoic Sediments of the Yushe Basin, Shanxi Province, China

Neil D. Opdyke, Kainian Huang, and R. H. Tedford

Abstract Twenty-four sections in the Yuncu subbasin of the Yushe Basin complex were sampled for magnetostratigraphy. A clear pattern of resolved magnetozones indicates the distinctive pattern of Late Miocene Chron C3An for the oldest part of the sequence, the Mahui Formation. The Gaozhuang Formation is dominantly reversely magnetized and correlated with the Gilbert Chron. The superposed Mazegou Formation represents Gauss Chron, but truncated magnetozones indicate a basal hiatus of about half a million years. These two formations are therefore Pliocene in age. The overlying reversely magnetized Haiyan Formation is correlated with Chron C2r, the early reversed subchron of the Matuyama, Early Pleistocene in age. Four sections in the Tancun subbasin record late Chron C3An and C3r (the latter being early Gilbert Chron). Deposition in both subbasins, near the trunk of the Zhuozhang River, began in the Late Miocene, antedating the commencement of sedimentation in the Zhangcun subbasin. Age control, initially provided by vertebrate fossils, is refined by the magnetostratigraphy, which provides precision for dating the Yushe succession of fossiliferous localities.

Keywords Paleomagnetism • Late Neogene • Yushe Basin • Yuncu subbasin • Zhangcun subbasin • Biostratigraphy

An erratum to this chapter is available at
10.1007/978-90-481-8714-0_6

N. D. Opdyke (✉) · K. Huang
Department of Geological Sciences, University of Florida,
Gainesville, FL 32611, USA
e-mail: drno@ufl.edu

K. Huang
e-mail: knhuang@ufl.edu

R. H. Tedford
Formerly Division of Paleontology, American Museum
of Natural History, Central Park West at 79 St, New York,
NY 10024, USA

4.1 Introduction

The late Cenozoic sediments of the Yushe Basin have been famous for many years for their abundant vertebrate fossil remains. Many fossils from the sediments were accumulated in a number of institutions in China, and sent to the University of Uppsala, Sweden, and the American Museum of Natural History in New York, USA, in the 1920s and 1930s. The present initiative, utilizing a combined stratigraphic and paleontological approach was therefore initiated in 1987 as a joint program between the Institute of Vertebrate Paleontology and Paleoanthropology (IVPP), Academia Sinica, Beijing, and the American Museum of Natural History in New York. The value of magnetic stratigraphy in studies of terrestrial biostratigraphy has been clear for several decades (Johnson et al. 1975, was a key landmark publication), since such studies enable the fossil sites to be dated and correlated globally. Therefore a magnetic stratigraphy component was added to the biostratigraphic investigation of the Yushe Basin sediments. This part of the study was based at the Department of Geological Sciences at the University of Florida.

4.2 Previous Studies

The Yushe Basin is an intermountain basin occupying the center of a synclinal structure formed in red–purple Triassic sandstone (Fig. 4.1). The fluviatile, lacustrine and loessic sediments of the late Neogene lie unconformably on the Triassic bedrock and have been tilted gently toward the northwest at angles usually less than 10°. The sediments are exposed in a series of subbasins situated between Yushe and Wuxiang towns. The Yuncu subbasin is the focus of this investigation and was chosen because of its long section and numerous fossil localities.

There have been two previous paleomagnetic investigations in Yushe sediments. The first was published by Cao et al. (1985) on the sediments of the Zhangcun subbasin,

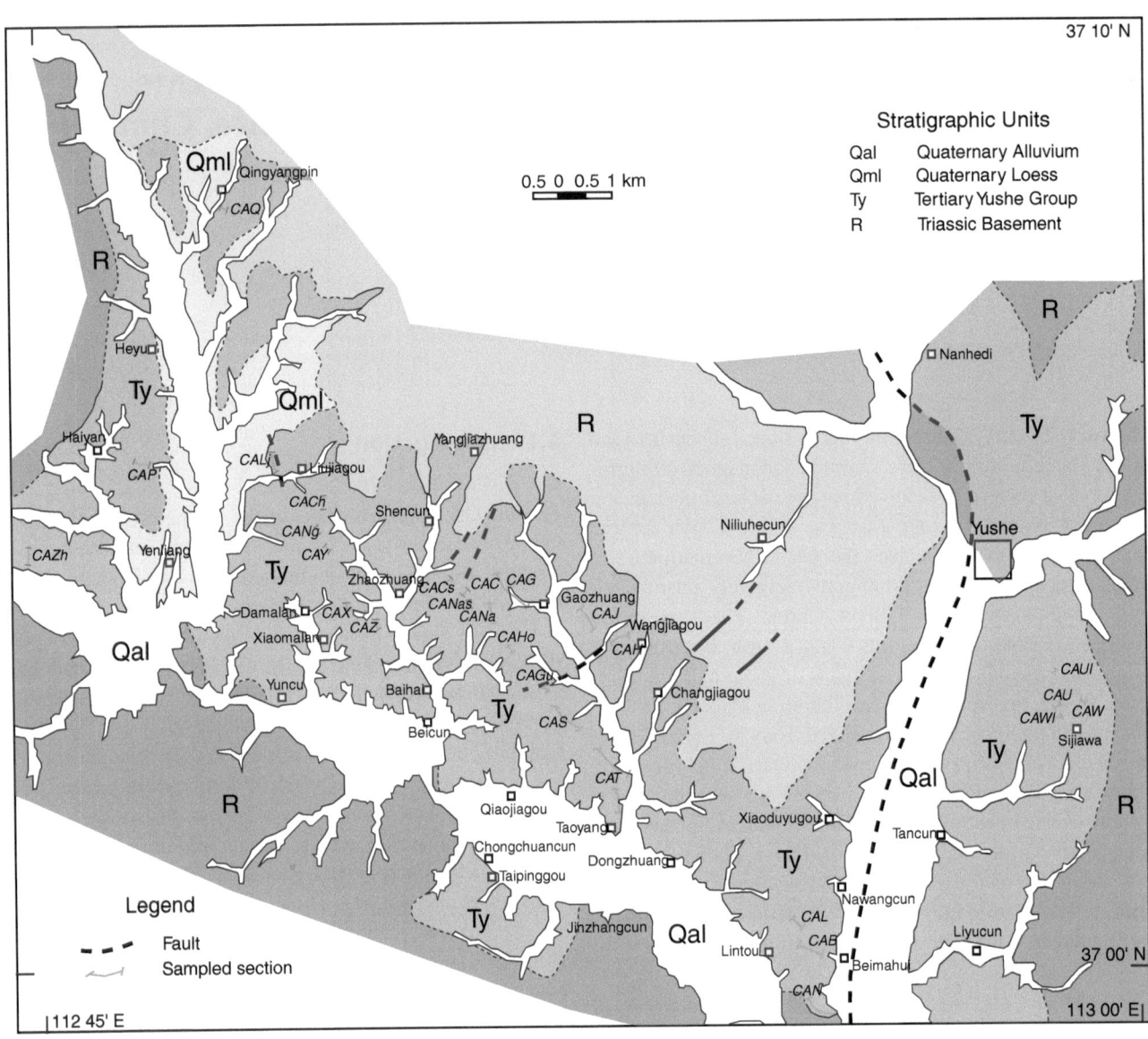

Fig. 4.1 Map of the Yuncu subbasin of the Yushe Basin (see Fig. 3.14). Sections are indicated by *gray irregular lines* with letter designations (see Fig. 3.4). Faults are indicated by *dashed black lines*

which lies within the Yushe Basin, but 5 km south of the Yuncu subbasin (the focus of the present investigation). This early study was followed in 1991 by the publication of preliminary results from the present study by Tedford et al. (1991). The stratigraphy and magnetostratigraphy of the Zhangcun subbasin was restudied by Shi Ning (1994), who published his revision several years later. That study provided the magnetic stratigraphy of a stratigraphic section 500 m in thickness and correlated to the lower Matuyama, Gauss, and most of the Gilbert Chron, which represents all of Pliocene time (Berggren et al. 1995). The correlation of the Zhangcun section with the results of the present study of the Yuncu subbasin will be discussed below.

4.3 Sampling and Laboratory Analysis

The paleomagnetic section in the Zhangcun subbasin was sampled along what is essentially a continuously exposed section. However, no single, continuously exposed litho-stratigraphic section is present within the Yuncu subbasin that contains the complete sedimentary sequence, therefore a series of short lithostratigraphic sections (Fig. 4.1) were measured and correlated by lithostratigraphic horizons, and the correlation was later assisted by magnetostratigraphy. The lithostratigraphy of the Yuncu subbasin is presented in this volume (Chap. 3). The lithostratigraphy for this study was compiled from a series of twenty-one short sections, the longest of which is 200 m. The paleomagnetic sites were sampled by hand; after digging with a pick to expose a fresh surface, a vertical or horizontal face was created on the sediments using a hand plane. An arrow was drawn on this surface using a pencil and oriented with a brunton compass. Three oriented samples were taken from each site. These sites were sampled from fine grained sediments and were spaced at approximately 5 m intervals if possible. The sites were then placed into the measured lithologic sections. The composite section is 800 m thick and has been divided into four formations, the Mahui, Gaozhuang, Mazegou and Haiyan formations (Qiu et al. 1987), which are thoroughly reviewed in Chap. 3 of this volume. Four unconformities recognized in the field break the sequence: at the base of the section on the Triassic sandstone and between each of the formations.

After the samples were collected, they were fashioned into cubes 2.4 cm on each side in the United States. The direction of magnetization was measured in a 2G cryogenic magnetometer. Previous experience has shown that progressive partial thermal demagnetization is the most effective procedure in removing magnetic overprints. All samples were therefore thermally demagnetized to temperatures of up to 670 °C or until the remnant vector was destroyed. Figure 4.2 gives the results from six samples from the Yushe Basin. In some samples the demagnetograms show essentially univectorial trends in both normal and reverse samples as shown in samples d, e, and f of Fig. 4.2. However, some samples have a large normal overprint which can be clearly recognized in reversed samples (e.g., Fig. 4.2c). In some weakly magnetized samples a stable

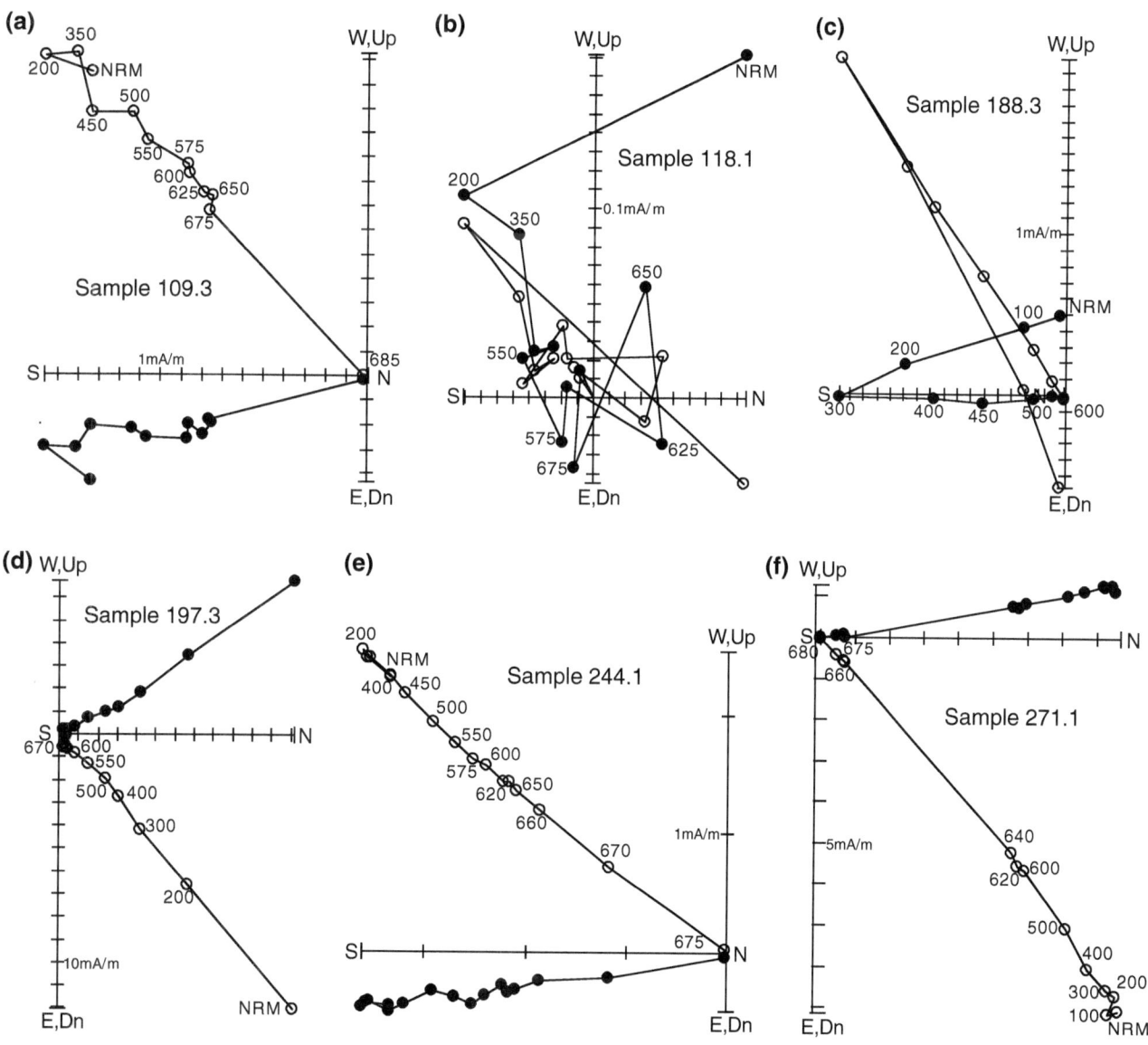

Fig. 4.2 Orthogonal projection of thermal demagnetization of samples from the Yushe Basin. *Open* and *closed symbols* indicate projection on the vertical and horizontal planes, respectively. The demagnetization temperatures are indicated in degrees centigrade and the intensity in milliampere per meter (mA/m)

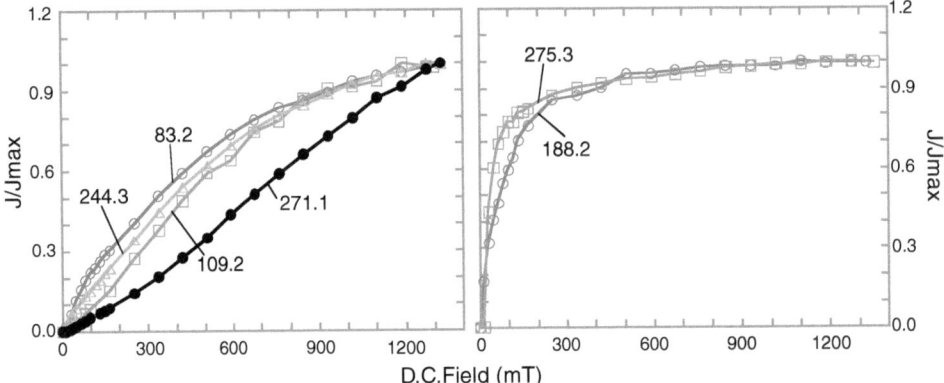

Fig. 4.3 Saturation experiment. Applied field in milli-Tesla (mT) plotted against induced magnetization (J/J max). The curves on the left are dominated by hematite since they never reach saturation, but the

samples on the right (188 and 275) show rapid increase in IRM intensity below 150 mT and slow-rise afterwards, suggesting the presence of magnetite as well as hematite

component is difficult to isolate, as in Fig. 4.2b; however, the magnetic polarity can be interpreted in some of these cases.

The unblocking behaviors illustrated in Fig. 4.2 indicate that hematite is the magnetic carrier in most sites (Fig. 4.2a, d–f) and magnetite dominates in some sites (Fig. 4.2c). See Tauxe et al. (1980) and Tauxe and Kent (1981) for unblocking behaviors characteristic of samples bearing hematite and magnetite and undergoing thermal demagnetization. Figure 4.3 presents IRM acquisition curves which again show that the remanence is carried by both magnetite and hematite. The curves on the left are dominated by hematite since they never reach saturation in the magnetic fields available. The samples on the right (188 and 275), however, show an initial rapid increase in IRM intensity below 150 mT and a clear slow-rise afterwards, suggesting the presence of both magnetite and hematite.

The characteristic direction of each sample was determined by line fitting techniques using the method of Kirschvink (1980). This direction was then averaged using Fisher (1953) statistics, and virtual geomagnetic pole (VGP) were calculated for each site with respect to overall pole position for all the site means.

Figure 4.4 shows the direction of magnetization for all samples plotted on a stereographic projection. It can be seen that the directions of magnetization form two well-grouped clusters, 180° apart, representing normal and reverse magnetized samples. The bedding dip is gentle so that no noticeable improvement in grouping is seen after correction for bedding tilt. The statistics of the northerly directed samples after tilt correction are $N = 87$, $R = 83.4$, $\kappa = 24.2$, $\alpha 95 = 3.1°$, $D = 1.1°$, $I = 34.7°$ and for the reversed samples are $N = 142$, $R = 134.8$, $\kappa = 19.5$, $\alpha 95 = 2.8°$, $D = 181.2°$,

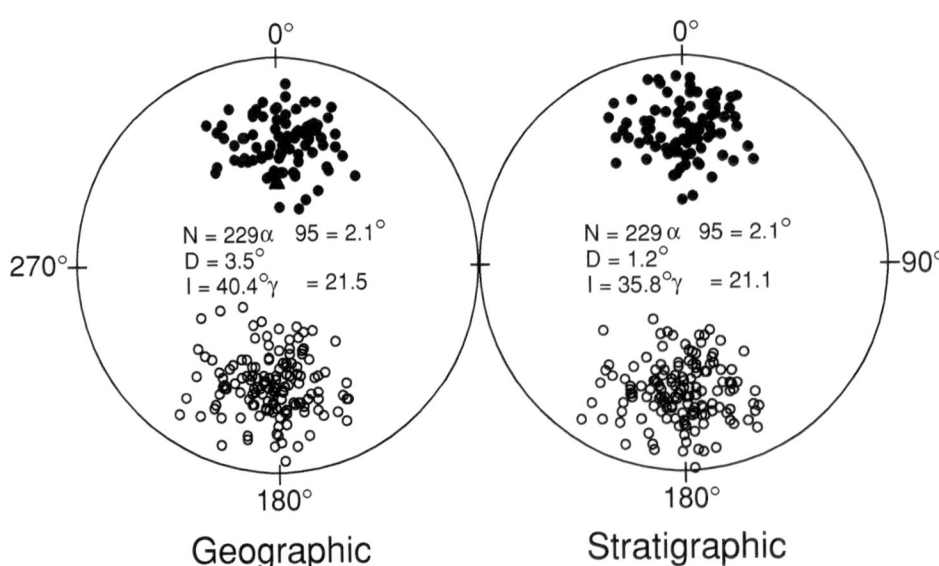

Fig. 4.4 Directions of magnetization plotted on a stereographic (Schmidt) net. Filled (open) circles indicate positive (negative) inclinations. N, D, I are number of samples, declination, inclination

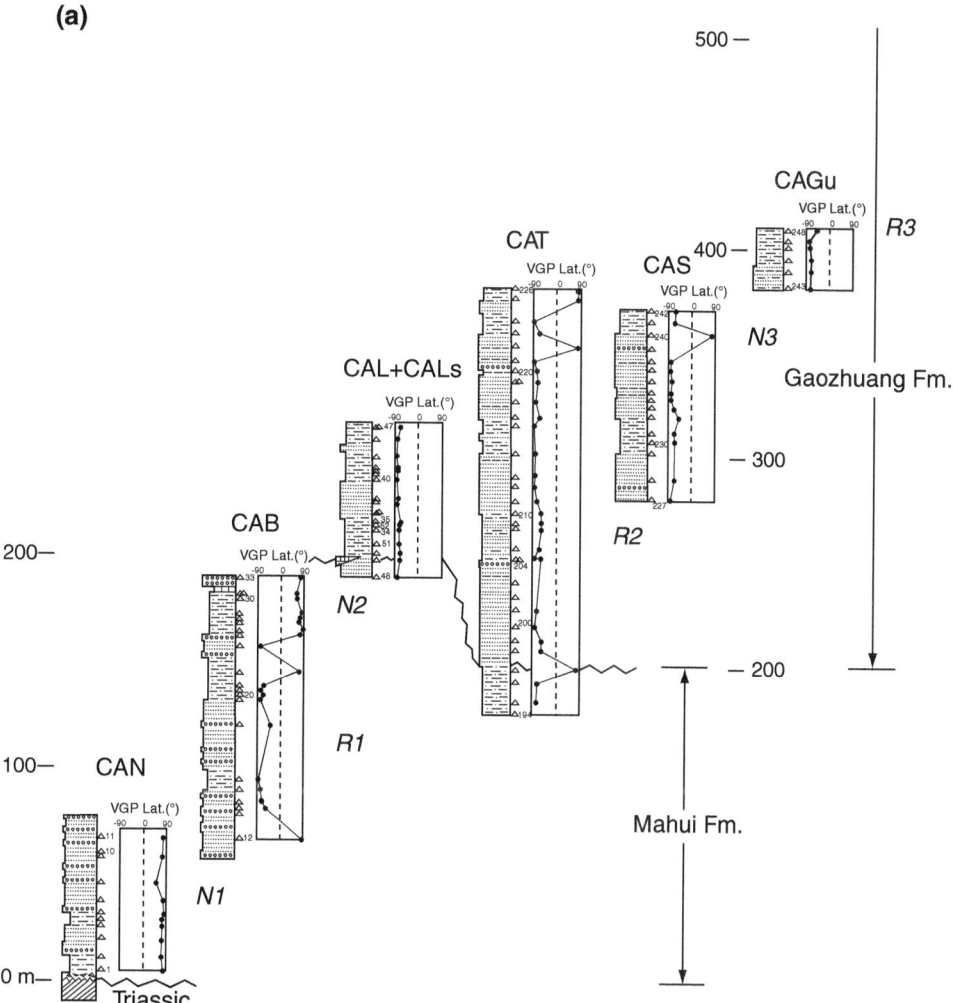

Fig. 4.5 The numbered triangles plotted on the lithostratigraphy indicate the level at which a site was collected and the VGP's are plotted at the resulting latitude derived from the declination (D) and inclination (I) of the site. The magnetozones beginning at the base of the stratigraphic section are designated N1, R1, ..., R8, from *bottom* to *top*. The initial correlation was done on lithostratigraphy (Fig. 3.4a). The correlation between sections CAN and CAB was made litho-stratigraphically in the field. Sections CAB and CAL are correlated using the unconformity at the top of the Mahui Formation. Section CAL is separated from sections CAT and CAS by about 5 km, but the Mahui-Gaozhuang contact could be traced to within 0.5 km of CAT at which point the rocks were faulted and the contact dropped down to the East. We assumed that the normal site at the base of section CAT is correlated to N2, most of which is removed on the Gaozhuang dis-conformity. We correlate the normal sites at the top of section CAT to the Sidufjal and Thvera subchrons. The correlation between sections CAT and CAS is supported by good lithologic criteria. Section CAGu does not overlap CAT and CAS, but we do not think the gap is great due to the low dips. **b** The *upper part* of the Gaozhuang Formation, continuing from **a**. Sections CAH to CAJ are tied together by a mar-ly—clay marker at the base and upper CAH and CAJ by another clay bed at that lies within the normal zone N4 (Nunivak). Another marker (the upper marl) lies in reverse polarity sediments. The upper part of the Gaozhung Formation lies within normal interval N5 (Cochiti). The basal Mazegou Formation at the top of the CAG section is reversed, late Gilbert. **c** Correlation of the Mazegou and Haiyan formation. Sections CAX and CAY do not overlap but the dips are low, so we believe that not much section is missing. We believe that R6 is present in CAY and CACh and represents a subchron within the Gauss Chron. There is an unconformity (*wavy line*) separating these sections from the Haiyan Formation of sections CAZh and CAP. Section CAQ is to the north (Fig. 4.1) and correlated lithologically

I = −36.5°. The α95s for the normal and reversed direction (inverted) have overlapping circles of confidence and pass the McFadden and McElhinny (1990) reversal test with an "A" classification. Therefore, an overall mean direction was cal-culated N = 229, R = 218.2, κ = 21.1, α95 = 2.1°, D = 1.2°, I = 35.8°, yielding a pole at latitude 72.7° North, longitude 289.1°. This pole position is far sided and results from the mean inclination being much shallower than the expected 45° for the latitude of the region of investigation. These sediments have an inclination error of 10°. It is possible that the error is caused by the presence of specular hematite in many of the samples as suggested previously by Tauxe et al. (1980) and Tauxe and Kent (1981) for Siwalik (Pakistan) sediments. No tectonic rotation is observed in these sediments.

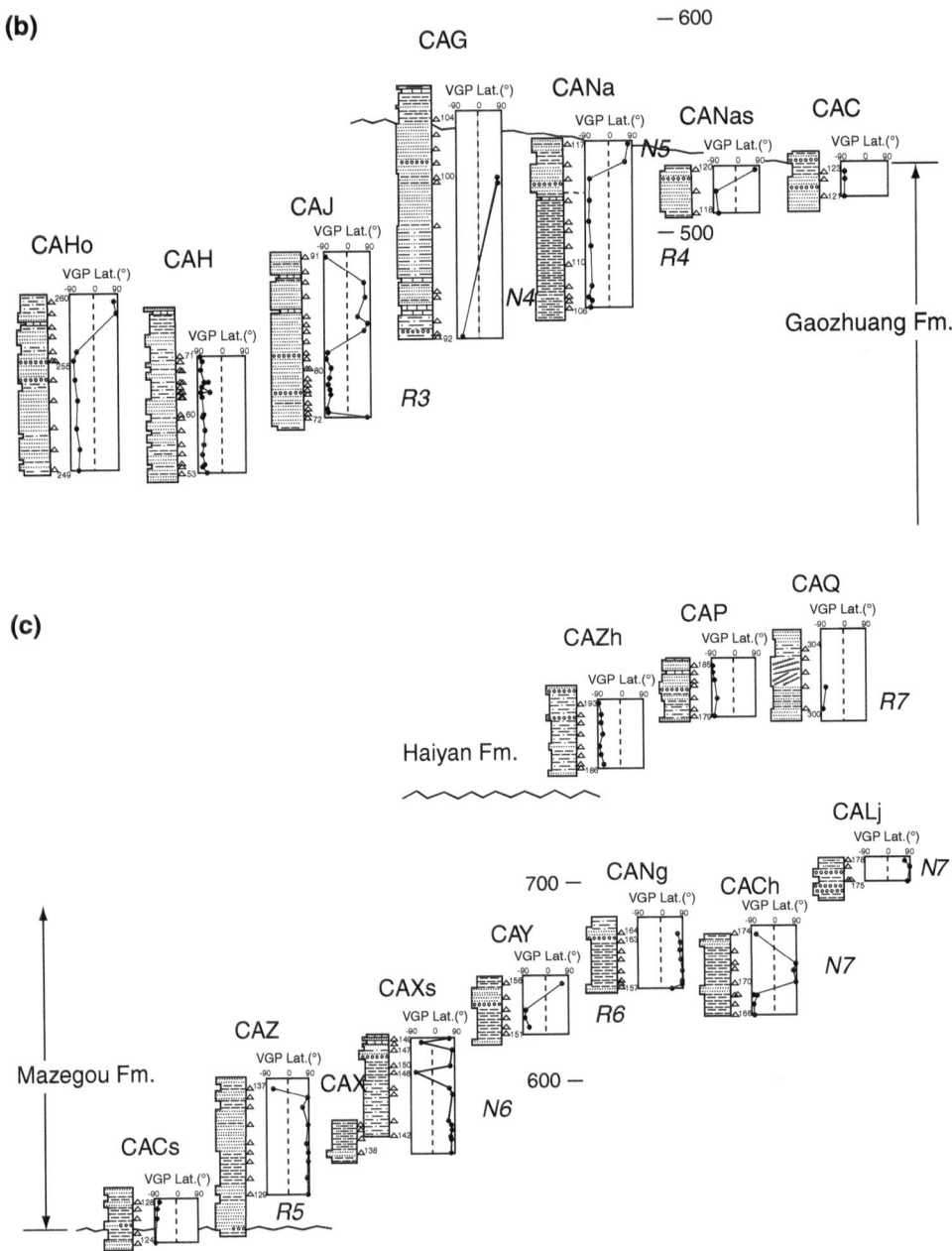

Fig. 4.5 (continued)

4.4 Magnetic Stratigraphy

VGP latitudes were placed in their correct stratigraphic position in the measured sections and a total of twenty magnetozones was established (Fig. 4.5a–c). These magnetozones were combined into a composite section with an aggregate thickness exceeding 800 m. The task of identifying these magnetozones against the magnetic polarity time scale (MPTS) proceeds by matching the observed pattern, with thicknesses approximating temporal durations, to the MPTS based, first, on a correlation informed by biochronological evidence. A similar approach was taken by Johnson et al. (1975). This correlation must have an independent dating framework as a starting point. In the case of Yushe Basin, fossils supply a range of ages for the series of formations from Late Miocene through the Pliocene, and since the sequence is capped by Late Pleistocene loess, its upper limit is the present Brunhes long normal (chron C1n).

The oldest sediments in this basin are known to be of Late Miocene age and the youngest formation is Early Pleistocene based on vertebrate paleontologic data (Qiu 1987; Qiu et al. 1987). The pattern of the magnetozones is rather distinctive.

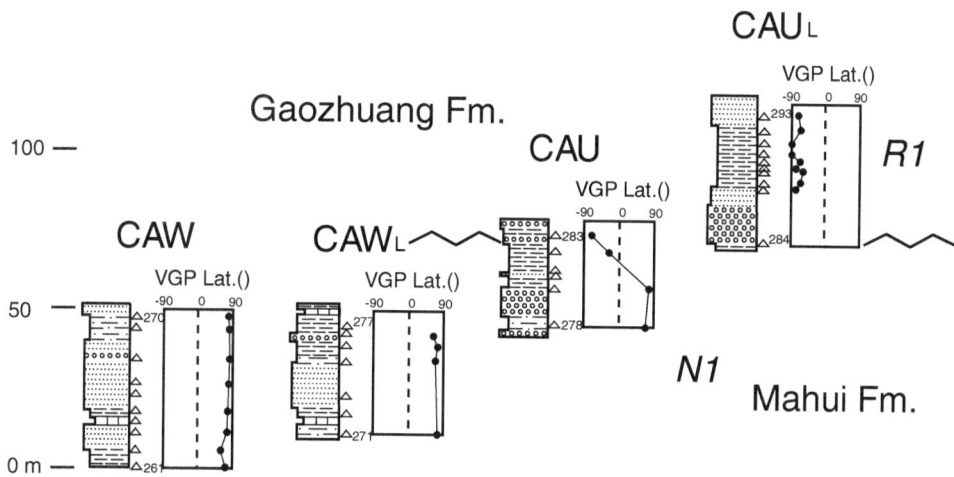

Fig. 4.6 Short sections in the northern Tancun subbasin with the top reversed and the bottom normal, correlating the base to the Mahui Formation, and upper levels to the lower Gaozhaung Formation (late Chron C3A and lower Gilbert). The unconformity (*wavy line*) is contained in reversely magnetized sediments

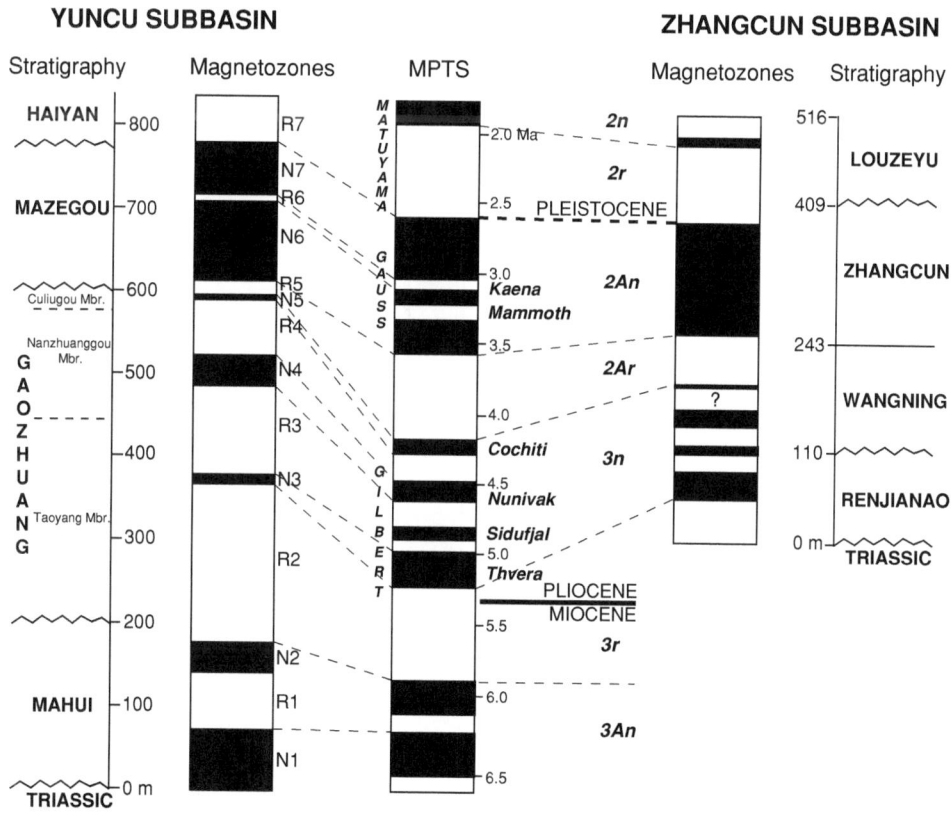

Fig. 4.7 Yushe Basin magnetostratigraphy. Correlation of the sections (m) in Yuncu subbasin (*left*) and the Zhangcun subbasin (*right*) to the Magnetic Polarity Time Scale (MPTS) (*middle*, Cande and Kent 1995), based on magnetic stratigraphy and vertebrate paleontology. Formation names are indicated for both subbasins, and most are bounded by unconformities (*wavy lines*). Member names for the Gaozhuang Formation of Yuncu subbasin are shown in smaller type, with *dashed lines* for their contacts. The alpha-numeric system of chron names is indicated in the *middle*, as are classical chron and subchron names (Gauss, Kaena, Thvera, etc.). *Light dashes* join observed transitions to proposed MPTS correlations. The Miocene–Pliocene limit is shown and the base of the [Pleistocene] is approximated by the chron C2An-C2r transition, which is a heavier *dashed line* than other *tie lines*

Fig. 4.8 **a**, **b** Composite Yuncu subbasin lithology with observed magnetostratigraphy and correlation to the GPTS of Cande and Kent (1995). Lithostratigraphic units are plotted with unconformities indicated by *wavy lines.* *Vertical arrows* in the GPTS column indicate location and duration of intra-Yushe hiatuses. This figure provides the framework for biostratigraphy developed in subsequent volumes of this Yushe series, and the YS fossil localities are plotted next to the composite lithological column. On the *right*, selected historical localities are located, some approximately indicated by a vertical range based on field evidence and statements by surviving relatives of the "dragon-bone hunters". These localities include results of IVPP fieldwork of the 1950s, Licent quarries (L), and collections made for Childs Frick (F; 1932-37). The two parts of the figure overlap, and are reproduced at a size large enough to resolve locations of stratigraphic levels

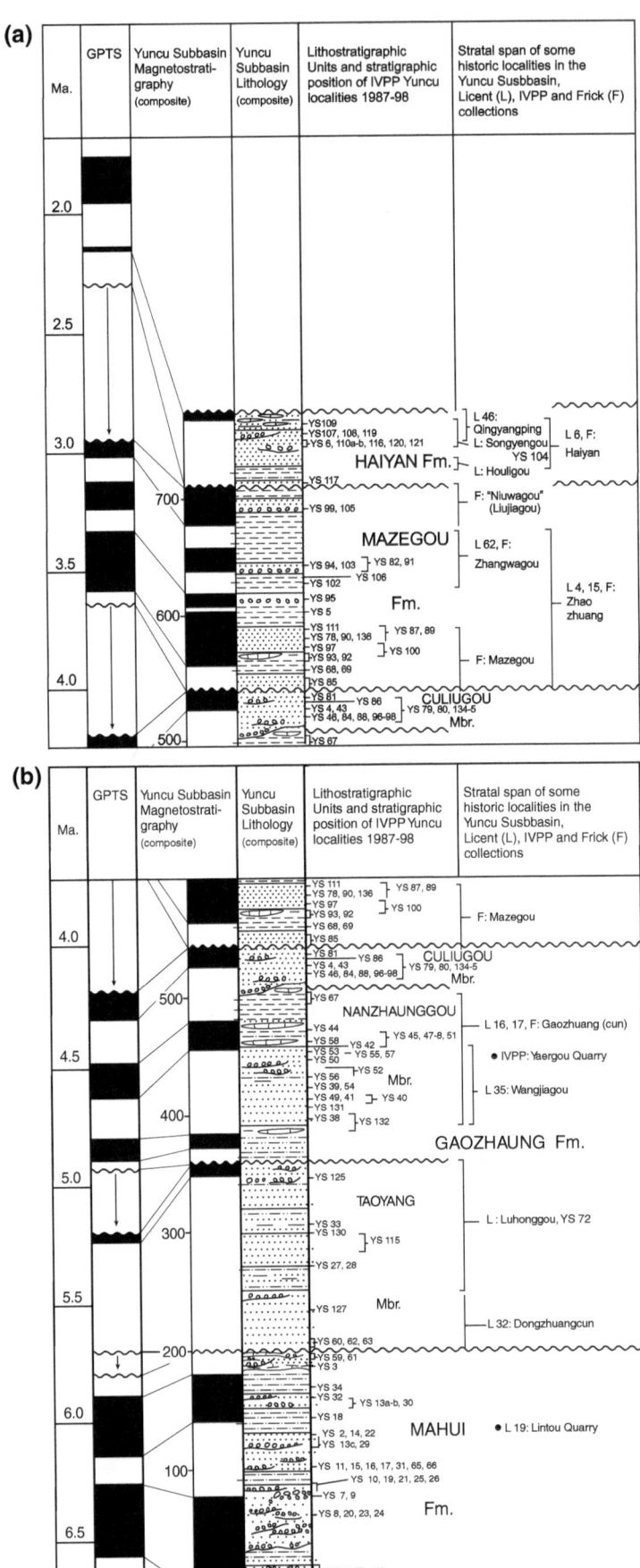

The Mahui Formation is dominated by a distinctive NRM (natural remanent magnetization) pattern, with two normally magnetized zones, while the Gaozhuang Formation is dominantly reversed with three short normal intervals (Fig. 4.5a, b). The Mazegou Formation is dominantly normal while the Haiyan Formation is reversed (Fig. 4.5c). Considering the time constraints imposed by the biochronology (Qiu et al. 1987; Tedford et al. 1991) the most likely interpretation of the magnetic stratigraphy is that the formations represent from bottom to top Chron C3An, followed by dominantly reversely magnetized sediment representing chrons C3r to C2Ar, normally magnetized C2An, and reversed C2r. The classical chron terminology lends well to the Yushe sequence: the Gaozhuang Formation was deposited during Gilbert Chron, the Mazegou Formation during Gauss Chron, and the Haiyan Formation during early Matuyama Chron. In terms of epoch boundaries, the Miocene–Pliocene boundary occurs high in the Taoyang Member of the Gaozhuang Formation, and the Haiyan Formation is Early Pleistocene in age. The formations are separated by unconformities with hiatuses of varying extent.

In 1991, our field work established the magnetostratigraphy for short sections totaling just over 100 m in the Tancun subbasin, south of the town of Yushe. As noted in Chap. 3, the composite (Fig. 4.6) represents upper levels of the Mahui Formation and the lower part of the Gaozhuang Formation. The normal magnetozone overlain by reversely magnetized sediment correlates to late Chron C3An and the subsequent Chron C3r, the older part of the Gilbert Chron.

The correlation proposed for the Yuncu subbasin and its local stratigraphy is presented on the left in Fig. 4.7. The biggest mismatch to the MPTS is within the Gilbert Chron. The normal magnetozones N3, N4 and N5 are not robustly correlated to the MPTS and undetected unconformities must be present in the Gaozhuang Formation. Magnetozone N4 is correlated to the Nunivak based on thickness above the Thvera, so the Sidufjal, in this interpretation, would be missing. Nevertheless, we feel that the overall interpretation presented here is reasonable, and we recognize the Gaozhuang and Mazegou formations as representing Gilbert and Gauss chrons.

The magnetic stratigraphy published by Shi (1994) from the adjacent Zhangcun subbasin (on the right in Fig. 4.7, with its local stratigraphic units) appears on first sight to give a complete match to the MPTS for the subchrons in the Gilbert. However, the data quality for the lower part of the Wangning Formation is poor, and this is the part of the section that correlates to the Sidufjal and Nunivak subchrons. There is no coherence between the declination and inclination values for this part of the sequence. A tentative correlation between the two subbasins is given in Fig. 4.7. Whereas the Yuncu subbasin records both normal magnetozones of Chron C3nAn, the Zhangcun subbasin does not,

and therefore the initiation of Yushe Group sedimentation began later there than in the Yuncu area.

One of the key project goals is placement of fossil localities unambiguously into the basin stratigraphy with dating from a well resolved paleomagnetic correlation. This effort provides the framework for Yushe biostratigraphy and allows testing the observed pattern by comparison with biostratigraphy from other basins. Therefore, the magnetostratigraphy is a unifying feature for all of our paleontological studies and for the synthesis of North China evolutionary biology as preserved in Yushe Basin. We placed our fossil localities (YS prefix) in the stratigraphic sections and determined as precisely as possible the stratigraphic positions of historical collections made by Licent, by collectors for Frick and the American Museum of Natural History, and by field parties of the Institute of Vertebrate Paleontology and Paleoanthropology. The function of Fig. 4.8 is biostratigraphy: to adapt the proposed magnetostratigraphic correlation of Fig. 4.7 to present the composite lithology and key fossil sites of Yushe on a time scale. The time scale (Ma, for megaannum, on the left) derives from the Cande and Kent (1995) GPTS. The biostratigraphic and systematic studies on Yushe paleontological collections developed in subsequent volumes of this Springer series refer to this chronology. Biostratigraphic data for YS localities represent field occurrences verified by members of SAYP, the Sino-American Yushe Project.

4.5 Conclusion

(1) The magnetic stratigraphy of the Yuncu subbasin ranges from chron C3N, at about 6.5 Ma, into chron C2N, probably at about 2.0 Ma or younger. (2) Correlation within the Gilbert Chron is inhibited by unconformities, which apparently cut out some subchrons. (3) The magnetic data are excellent and show an inclination error of about 10° with no tectonic rotation. (4) The base of the Yushe Group in the Yuncu subbasin appears to be older than that from the adjacent Zhangcun subbasin.

Acknowledgments The writers thank the personnel (then students) from IVPP and from the Xi'an Laboratory of Loess and Quaternary Geology, Xi'an, Shaanxi Province, for help in collection of samples reported here. In particular, we are indebted to Donghuai Sun, Xiao Jule and Hong Wang.

References

Berggren, W. A., Kent, D. V., Swisher, C. C. III, & Aubry, M.-P. (1995). A revised Cenozoic geochronology and chronostratigraphy. In W. A. Berggren, D. V. Kent, M.-P. Aubry & J. Hardenbol (Eds.), *Geochronology, time scales and global stratigraphic correlation* (pp. 129–212). Tulsa: (SEPM) Society for Sedimentary Geology.

Cande, S. C., & Kent, D. V. (1995). Revised calibration of the geomagnetic polarity timescale for the Late Cretaceous and Cenozoic. *Journal of Geophysical Research, 100*, 6093–6095.

Cao, Z.-Y., Xing, L., & Yu, Q. (1985). The age and boundary of magnetic strata of the Yushe Formation. *Bulletin of the Institute of Geomechanics CAGS, 6*, 143–153.

Fisher, R. A. (1953). Dispersion on a sphere. *Proceedings of the Royal Society London, A217*, 295–305.

Johnson, N. M., Opdyke, N. D., & Lindsay, E. H. (1975). Magnetic polarity stratigraphy of Pliocene Pleistocene terrestrial deposits and vertebrate faunas, San Pedro Valley Arizona. *Geological Society of America Bulletin, 86*, 5–12.

Kirschvink, J. L. (1980). The least squares lines and plane analysis of paleomagnetic data. *Journal of the Royal Astronomical Society, 62*, 699–718.

McFadden, P. L., & McElhinny, M. W. (1990). Classification of the reversal test in Paleomagnetism. *Geophysical Journal International, 103*, 725–729.

Qiu, Z.-X. (1987). Die Hyaeniden aus dem Ruscinium und Villafranchium Chinas. *Münchener Geowissenschaftliche Abhandlungen, A9*, 1–110.

Qiu, Z.-X., Huang, W.-L., & Guo, Z.-H. (1987). The Chinese hipparionine fossils. *Palaeontologia Sinica, New Series, C25*, 1–243 (in Chinese with English summary).

Shi, N. (1994). The Late Cenozoic stratigraphy, chronology, palynology and environmental development in the Yushe Basin, North China. *Striae, 36*, 1–90.

Tauxe, L., & Kent, D. V. (1981). Properties of a detrital remanence carried by hematite from study of modern river deposits and laboratory redeposition experiments. *Geophysical Journal of the Royal Astronomical Society, 77*, 543–561.

Tauxe, L., Kent, D. V., & Opdyke, N. D. (1980). Magnetic components contributing to the NRM of Middle Siwalik red beds. *Earth and Planetary Science Letters, 47*, 279–284.

Tedford, R. H., Flynn, L. J., Qiu, Z.-X., Opdyke, N. D., & Downs, W. R. (1991). Yushe Basin, China: Paleomagnetically calibrated mammalian biostratigraphic standard for the Late Neogene of eastern Asia. *Journal of Vertebrate Paleontology, 11*(4), 519–526.

Chapter 5
Biostratigraphy and the Yushe Basin

Lawrence J. Flynn and Zhan-Xiang Qiu

Abstract Volume I of *Late Cenozoic Yushe Basin, Shanxi Province, China: Geology and Fossil Mammals* presents the physical setting of Yushe Basin on the eastern edge of the Loess Plateau in North China, and documents the stratigraphy and paleomagnetic dating of the Yushe Group. The importance of Yushe Basin in the development of modern vertebrate paleontology in China is explored through historical records. Yushe Basin remains the single best relatively continuous Late Neogene sequence in North China on which to base Pliocene biostratigraphy. Its multiple fossil horizons spanning the Late Miocene to Early Pleistocene permit definition of biochrons, examination of faunal turnover, estimation of local residence durations at the species level, and access to an important source of data on biogeographical distributions of Late Neogene mammals.

Keywords Yushe Basin • Biostratigraphy • Astronomically Tuned Neogene Time Scale • Biochronology • Late Neogene

5.1 Biochrons, Magnetostratigraphy, and Biostratigraphy in Yushe Basin

The Yushe Basin, in particular the Yuncu subbasin, preserves a wealth of Late Neogene paleontological data, especially vertebrate fossils. The geological and magnetostratigraphic work done under the Sino-American Yushe Project field program (SAYP, 1987–1998) confirms the ages encompassed by strata of the Yushe formations to be Late Miocene through the Pliocene, to Early Pleistocene. It therefore contains a rich fossil data source that documents knowledge of terrestrial faunas of North China and represents most of northern Asia. Yushe deposits and faunas span considerable time, over 6 million years. The Mahui Formation is Miocene in age, spanning about 6.7–5.8 Ma on the Gradstein et al. (2004) time scale (Fig. 5.1). This unit is best developed in Yuncu subbasin, but the Gaozhuang Formation is more widespread and represented in most subbasins. The base of the Gaozhuang Formation is Late Miocene in age, and the top is medial Pliocene, about 4.2 Ma. The succeeding Mazegou Formation is bounded by hiatuses and spans about 3.7–2.8 Ma. The overlying Haiyan Formation represents a short interval in the Early Pleistocene. Successive loess blankets span much of the Early Pleistocene.

In faunal terms (Qiu et al. 2013) the Asian land mammal ages represented by Yushe faunas are the Baodean Land Mammal Age (Late Miocene), Gaozhuangian and Mazegoan Land Mammal Ages (Pliocene), and the Nihewanian Land Mammal Age (Early Pleistocene). The Pliocene Gaozhuangian and Mazegouan are named for successive fossiliferous formations from Yuncu subbasin: the older Gaozhuang Formation and superposed Mazegou Formation. A key purpose for subsequent volumes of the Springer Yushe series is to characterize the faunal components of Gaozhuang and Mazegou assemblages to document the representation of these Ages in Yushe Basin.

The stratigraphic framework built by Tedford, Qiu, and Ye (Chap. 3) allowed development of a refined magnetostratigraphy for Yuncu subbasin. Opdyke, Huang, and Tedford (Chap. 4) resolved Yushe magnetozones, identifying them against the Geomagnetic Polarity Time Scale (GPTS). They showed that the Mahui Formation and its fossils correlate to Chron C3An, equivalent to younger levels of the classic Baode Formation (which spans approximately 7.2–5.4 Ma; Zhu et al. 2008). Most vertebrates at Baode derive from

L. J. Flynn (✉)
Department of Human Evolutionary Biology and Peabody Museum of Archaeology and Ethnology, Harvard University, Cambridge, MA 02138, USA
e-mail: ljflynn@fas.harvard.edu

Z.-X. Qiu
Laboratory of Paleomammalogy, Institute of Vertebrate Paleontology and Paleoanthropology, Chinese Academy of Sciences, Xizhimenwai Ave., 142, Beijing, 100044, People's Republic of China
e-mail: qiuzhanxiang@ivpp.ac.cn

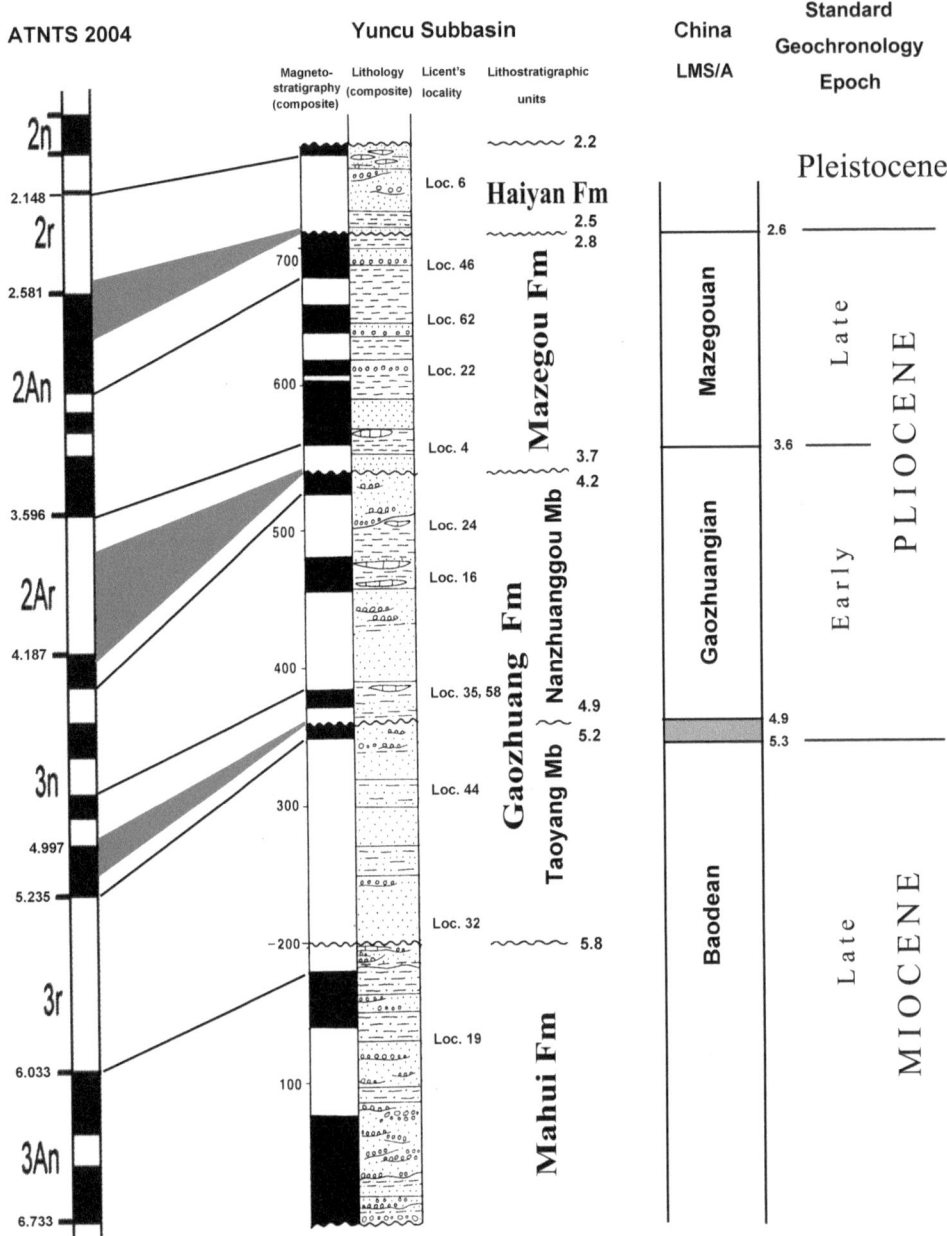

Fig. 5.1 Composite stratigraphy and magnetostratigraphy for the Yuncu subbasin of Yushe Basin. In the middle of the figure is the observed stratigraphy and stratigraphic units, with observed magnetostratigraphy and placement of key fossil localities of Emile Licent (Loc. number). The magnetozones are correlated toward the left to the Astronomically Tuned Neogene Time Scale, ATNTS 2004, developed by Lourens et al. (2004). Tie point correlations of magnetic transitions are indicated, and missing time is shown by gray shading. On the *right*, the successive Late Neogene Chinese Land Mammal Stage/Ages as represented in Yushe Basin are arranged with approximate age boundaries indicated. The Early Pleistocene Nihewanian Stage/Age is represented by faunas of the Haiyan Formation and follows the Mazegouan Stage/Age. Epoch boundaries are also indicated. The Miocene-Pliocene boundary preceeds the Thvera event at about 5.3 Ma; the lower part of the Gaozhuang Formation is therefore late Baodean and Miocene in age. The Pliocene–Pleistocene boundary is at the *top* of chron 2An (*top* of Gauss), at about 2.6 Ma (*heavy dashed line*), and falls in the hiatus between the Mazegou and Haiyan formations. This figure is a modified version of Fig. 1.6 in Qiu et al. (2013)

sediment older than 6.4 Ma in age (Zhu et al. 2008), so the Mahui Formation faunas of about 6 Ma importantly present an excellent record of late Baodean time. The exception for Baode is Locality 30, which is late Baodean, about 5.7 Ma according to Kaakinen et al. (2013). Opdyke and colleagues showed generally that Gaozhuang levels are identified with magnetic chrons C3r to C2Ar, the Gilbert magnetozones, and that Mazegou deposits correlate with C2An, the Gauss Chron. Most of the Taoyang Member of the Gaozhuang Formation correlates to chron C3r, and is therefore latest Miocene in age.

Reversely magnetized sediment of the Haiyan Formation represents part of chron C2r, pre-Olduvai Matuyama Chron, Early Pleistocene in current usage (see Chap. 1).

Opdyke and colleagues built a case for inter-subbasin variation in mode of deposition. In addition to lithological differences widely noted (e.g., Chap. 3), they argue that Yushe Group sedimentation began somewhat earlier in the Yuncu-Tancun area than elsewhere in Yushe Basin. Some evidence (Sect. 3.7.1) suggests that Wuxiang Basin sedimentation farther down the trunk of the major drainage, the Zhuozhanghe, began in the Late Miocene, somewhat earlier than in Yushe Basin. The magnetostratigraphic framework (Fig. 4.8a, b) is referenced in the systematic treatments of the subsequent volumes of this Springer series.

Yushe Basin faunal assemblages do indeed embrace important times of critical transitions in the terrestrial vertebrate history of Asia: events at the end of the Miocene and the commencement of the Pleistocene. The quality of the basin complex that stands out beyond this is the number of superposed faunal levels comprised over this long interval; there are many. Biostratigraphic data are tightly controlled, especially for the newer collections of the last 60 years, and age resolution by means of magnetostratigraphy is on the scale of 10^5 years. For some time spans, every 100,000 year interval has faunal data. Exceptions are the hiatuses (Fig. 4.8a, b) and some depauperate strata, such as in the lower Mahui Formation and the lower Gaozhuang Formation, where there are few fossil localities (Fig. 4.8a, b). In addition, among the known fossil occurrences the data quality is not even. For example, Taoyang Member fossil sites are generally not as rich as overlying and underlying localities. Bearing these caveats in mind, however, the Yushe sequence biostratigraphy has been developed to the level that more penetrating paleobiological questions can be approached by the data set, than would be possible without good control (Flynn and Wang 1997). Examples include measuring longevities (residence times) of species in Yushe Basin, pinpointing times of introduction of exotic taxa, and measuring mode and degree of faunal turnover. Paleobiological questions of finer scale can be approached now through the Yushe data set, questions relevant for unraveling the terrestrial history of all of North China.

Tedford (1995) reflected on the growth of biostratigraphic work in China since the 1970s. This growth is accelerating in the twenty-first century. Tedford (1995) used Yushe biostratigraphy to illustrate an initial step in developing biochrons for the Late Neogene of North China and potentially for much of Asia. The process involves resolution of local biostratigraphy, with independent dating (as from paleomagnetic analysis), and comparison with dated biostratigraphies form other basins. By testing the homotaxis of key faunal components from different basins, concurrent range zones of mammalian taxa can be identified and examined for their geographical distribution. If widespread and repeatedly observed, index taxa may be recognized and used to establish biochrons, ultimately Asian Land Mammal Stages/Ages. This process is still maturing for Asian biochronology.

5.2 Yushe Basin Events and the ATNTS Time Scale

We draw reader attention to the fact that we have used various time scales over the last two decades, and it is the responsibility of researchers to take this into account when compiling data from older sources. Tedford et al. (1991) used the then-widely-cited Berggren et al. (1985) GPTS. Later publications including Chap. 4 herein have used and preferred the Cande and Kent (1995) time scale. Figure 5.1 makes use of the recently widely-endorsed Astronomically Tuned Neogene Time Scale (ATNTS) of Lourens et al. (2004) in the volume *A Geologic Time Scale* by Gradstein et al. (2004). Gradstein et al. (2004) is in turn outdated, and we can expect refinements in future time scales.

ATNTS 2004 yields somewhat different age estimates for stratigraphic boundaries than have been cited in the past. In comparison, the Cande and Kent (1995) time scale yields somewhat younger ages for older rock units. The bottom and top of the Mahui Formation are about 6.4 and 5.8 Ma on the Cande and Kent (1995) time scale; there is a slight hiatus at the base of the Gaozhuang Formation, and its top is about 4.2 Ma; the Mazegou Formation spans about 3.7–2.9 Ma, nearly the same on ATNTS 2004; the Haiyan Formation is about the same.

While time scales differ, each is robust internally and, as used for Yushe, fossil sites are plotted precisely relative to each other on the basis of observed stratigraphy. Ages for fossil occurrences can be calculated using a time scale, and despite uncertainty of absolute dates, ages relative to each other can be resolved to the scale of at least 10^5 years. For example, relative ages for the Licent localities of the Mazegou Formation (Fig. 5.1) can be resolved between themselves to probably less than 100,000 year, regardless of the actual age.

Figure 5.1 illustrates the time spans represented by Yuncu subbasin formations. Also indicated for the Gaozhuang Formation, are the durations of the Taoyang and Nanzhuanggou Members, two major depositional cycles described in Chap. 3. Figure 5.1 does not distinguish the thin Culiugou Member overlying the Nanzhuanggou Member because it is largely removed by erosion. One function of the figure is to relate the lithology to the hypothesized Gaozhuangian and Mazegouan land mammal stage/ages. The older Baodean age persists to near the Miocene–Pliocene boundary; it is not yet clear from biostratigraphy exactly where the Baodean-Gaozhuangian transition would fall (hence the shaded 5.3–4.9 interval). The Gaozhuangian age is based mainly on fossils from the Nanzhuanggou (plus Culiugou) Member. Mazegouan commences just above the bottom contact of the Mazegou Formation and lasts through the end of the Pliocene. The Early Pleistocene Nihewanian is named for the Nihewan Basin of Hebei, and is represented in Yushe Basin by *Equus* faunas of the Haiyan Formation.

The hiatuses in the Yushe stratigraphic record are in some cases evident in the field, but that between the Gaozhuang and Mazegou formations has emerged mainly because the long magnetostratigraphy can be matched to the GPTS; missing and truncated magnetozones become readily apparent. Opdyke's magnetic study (Chap. 4) allows estimation of hiatus durations. Hiatuses have been portrayed in various publications on Yushe stratigraphy of the last 20 years (see Tedford et al. 1991). Figure 5.1 also illustrates these by use of gray shading to indicate missing time.

Yushe Basin, located in eastern Shanxi Province, west of but near the Taihang Shan, occupies the eastern edge of the Loess Plateau. The Late Miocene fauna of Yushe may be compared with the Baode faunas 300 km to the northwest, and thus more inland in location, and more continental in paleoecology. Faunal analyses that follow in later volumes of this Springer series will explore comparisons in detail. Our preliminary impression is that Yushe presents a significantly more equable and probably moister environment for Baodean Age mammals, this based on somewhat different faunal composition. Therefore, one goal will be to study the paleobiology of the Yushe setting, which appears to represent a variant of typical North China Baodean assemblages. The same may hold for Pliocene faunas.

The Springer volumes to follow in this series will treat the fossil mammals of Yushe Basin in systematic fashion. We shall describe new material, as well as specimens from older collections that have not been studied. Volumes will focus on Orders of Mammalia as noted in Chap. 1. At the same time, we shall strive to highlight and summarize Late Neogene biotic changes during the last six million years that characterize past conditions for much of northern Asia. Events include the close of equable Pliocene conditions and the onset of Pleistocene climates, which set the stage for faunal turnover and introduction of a new faunal element, *Homo*.

References

Berggren, W. A., Kent, D. V., Flynn, J. J., & Van Couvering, J. A. (1985). Cenozoic geochronology. *Geological Society of America Bulletin, 96*, 1407–1418.

Cande, S. C., & Kent, D. V. (1995). Revised calibration of the geomagnetic polarity timescale for the Late Cretaceous and Cenozoic. *Journal of Geophysical Research, 100*, 6093–6095.

Flynn, L. J., & Wang, B.-Y. (1997). Toward a denser biotic record for questions of finer scale. Proceedings 30th international geological congress. *VSP International Science Publications, 21*, 11–23.

Gradstein, F. M., Ogg, J. G., & Smith, A. G. (Eds.). (2004). *A geologic time scale 2004*. Cambridge: Cambridge University Press.

Kaakinen, A., Passey, B. H., Zhang, Z.-Q., Liu, L.-P., Pesonen, L. J., & Fortelius, M. (2013). Stratigraphy and paleoecology of the classical dragon bone localities of Baode County, Shanxi Province. In X. Wang, L. J. Flynn, & M. Fortelius (Eds.), *Fossil mammals of Asia: Neogene biostratigraphy and chronology* (pp. 29–90). New York: Columbia University Press.

Lourens, L., Hilgen, F., Shackelton, N. J., Laskar, J., & Wilson, D. (2004). The Neogene period. In F. M. Gradstein, J. G. Ogg, & A. G. Smith (Eds.), *A geologic time scale* (pp. 409–452). Cambridge: Cambridge University Press.

Qiu, Z.-X., Deng, T., Qiu, Z.-D., Li, C.-K., Zhang, Z.-Q., Wang, B.-Y., et al. (2013). Neogene land mammal ages of China. In X. Wang, L. J. Flynn, & M. Fortelius (Eds.), *Fossil mammals of Asia: Neogene biostratigraphy and chronology* (pp. 29–90). New York: Columbia University Press.

Tedford, R. H. (1995). Neogene mammalian biostratigraphy in China: Past, present, and future. *Vertebrata PalAsiatica, 33*(4), 272–289.

Tedford, R. H., Flynn, L. J., Qiu, Z.-X., Opdyke, N. D., & Downs, W. R. (1991). Yushe Basin, China: Paleomagnetically calibrated mammalian biostratigraphic standard for the Late Neogene of eastern Asia. *Journal of Vertebrate Paleontology, 11*(4), 519–526.

Zhu, Y.-M., Zhou, L.-P., Mo, D.-W., Kaakinen, A., Zhang, Z.-Q., & Fortelius, M. (2008). A new magnetostratigraphic framework for late Neogene Hipparion Red Clay in the eastern Loess Plateau of China. *Palaeogeography, Palaeoclimatology, Palaeoecology, 268*, 47–57.

Chapter 6
Erratum To: The Paleomagnetism and Magnetic Stratigraphy of the Late Cenozoic Sediments of the Yushe Basin, Shanxi Province, China

Neil D. Opdyke, Kainian Huang, and R. H. Tedford

Erratum To: The Paleomagnetism and Magnetic Stratigraphy of the Late Cenozoic Sediments of the Yushe Basin, Shanxi Province, China, DOI: 10.1007/978-90-481-8714-0_4

Figure 4.8 part figures **a** and **b** are placed on separate pages for clarity.

The online version of the original chapter can be found at
DOI: 10.1007/978-90-481-8714-0_4.

N. D. Opdyke (✉) · K. Huang
Department of Geological Sciences, University of Florida,
Gainesville, FL 32611, USA
e-mail: drno@ufl.edu

K. Huang
e-mail: knhuang@ufl.edu

R. H. Tedford
Formerly Division of Paleontology, American Museum
of Natural History, Central Park West at 79 St, New York,
NY 10024, USA

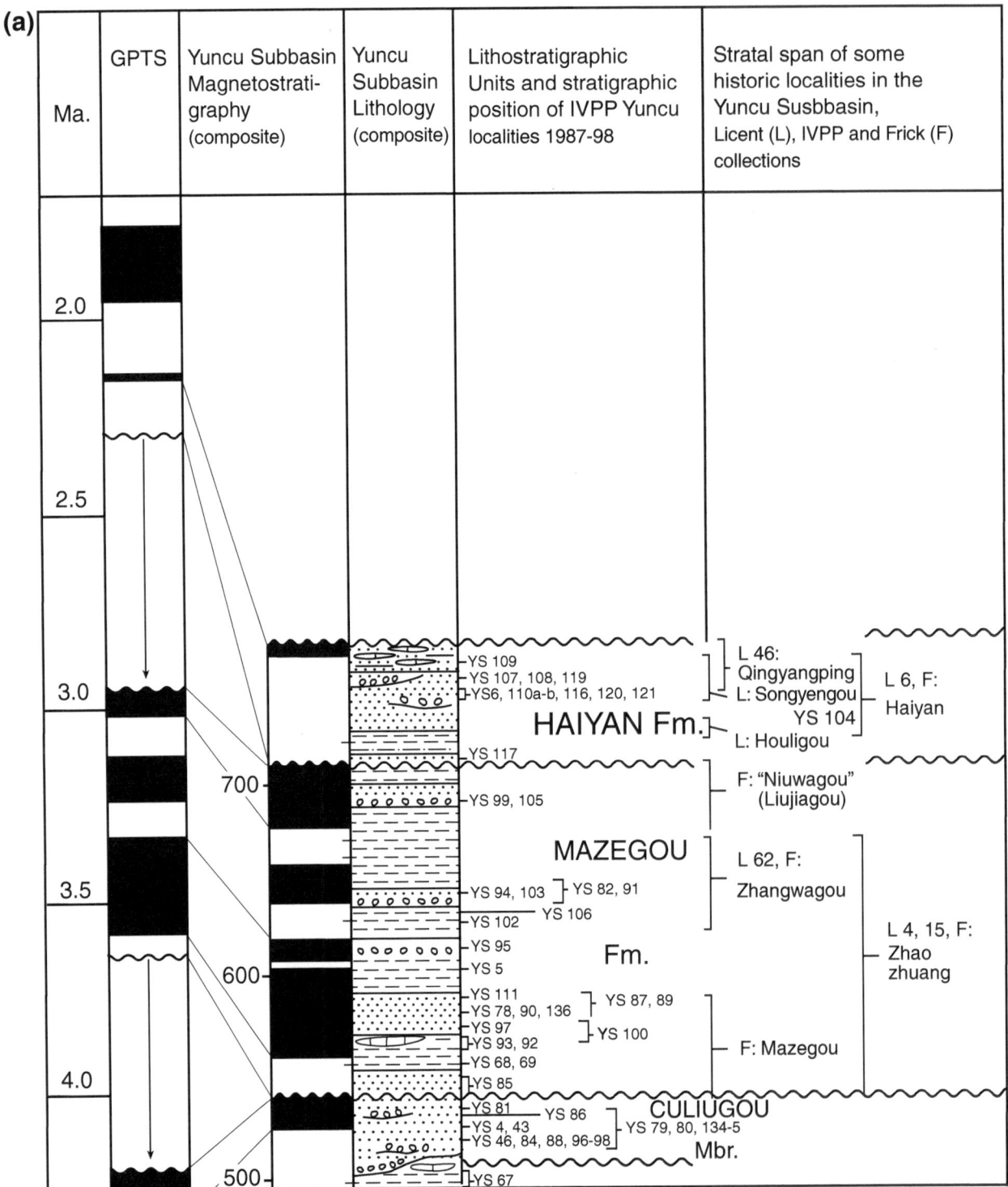

Fig. 4.8 a, b Composite Yuncu subbasin lithology with observed magnetostratigraphy and correlation to the GPTS of Cande and Kent (1995). Lithostratigraphic units are plotted with unconformities indicated by *wavy lines. Vertical arrows* in the GPTS column indicate location and duration of intra-Yushe hiatuses. This figure provides the framework for biostratigraphy developed in subsequent volumes of this Yushe series, and the YS fossil localities are plotted next to the composite lithological column. On the *right*, selected historical localities are located, some approximately indicated by a vertical range based on field evidence and statements by surviving relatives of the "dragon-bone hunters". These localities include results of IVPP fieldwork of the 1950s, Licent quarries (L), and collections made for Childs Frick (F; 1932-37). The two parts of the figure overlap and are reproduced at a size large enough to resolve locations of stratigraphic levels

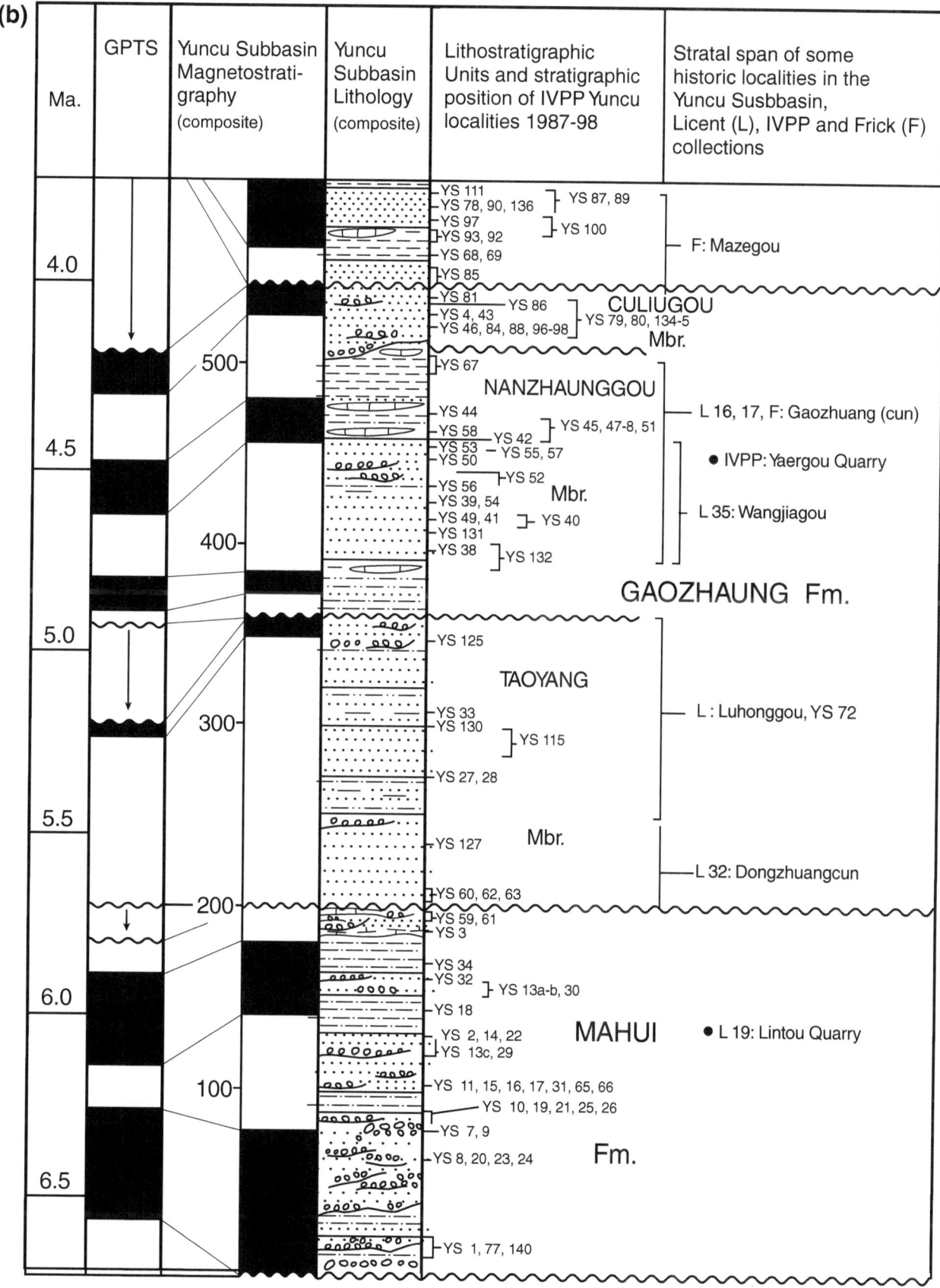

Fig. 4.8 (continued)

Appendix I
Andersson's Localities in Southern Shanxi Province

Geographic names using outdated Wade-Giles or other systems are in *italics*, their current usage in Pinyin Romanization System is followed in brackets (). *Note*: *NL* not located.

Andersson lok. name	Geographic location
32 *Hsiao-Fu-Tsun, Tung-Tai* (Xiaofucun, Dongtai, NL)	*Chi-Hsian* (Ji Xian, Fig. 2.1)
33 *Lung-Wang-Kou, Ma-Ti* (Longwanggou, Madi, NL)	*Hsiangning* (Xiangning, Fig. 2.1)
34 *Pantaopo* (Bandaopo, NL)	*Hsiangning* (Xiangning, Fig. 2.1)
41 *Hsiangning* (Xiangning, Fig. 2.1)	No exact locality
70 *Hsikoutsun, Chingkou* (Xigoucun, Qinggou)	6 km SSE of Wuxiang County seat
71 *Hsikoutsun, Hutzukou* (Xigoucun, ?)	6 km SSE of Wuxiang County seat
72 *Haojiazhuang, Xingwagou* (?Haojianao, ?)	?4 km S of Wuxiang County seat
73 *Tungtsun, Touchiaokou* (Dongcun, Doujiaogou)	1.1 km E of Wuxiang County seat
74 *Wangchiachi* (?Wangjiaji, NL)	*Wuhsiang* (Wuxiang)
75 *His-Kou-Tsun, Po-Cheng-Tzu* (Xigoucun, ?)	6 km SSE of Wuxiang County seat
77 *Hao-Chia-Po* (?Haojianao)	?4 km S of *Wuhsiang* (Wuxiang)
78 *Tsao-Tsun* (Caocun)	0.5 km S of Xigoucun, Wuxiang
80 *Nan-Ting, Hou-Tzu-Kou* (Nanting)	7 km S of Wuxiang County seat
81 *Ho-Chien-Nao* (Hejiannao)	4 km S of Wuxiang County seat
Wei-Chia-Yao, Chai-Kou (?Weijiayao, NL)	4.5 km SE of Wuxiang County seat
Yin-Chiao-Tsun, Hsi-Liang (Yinjiao, west crest)	Nihe Subbasin, Yushe, Fig. 2.3
Ch'iao-Chia-kou (Qiaojiagou)	Yuncu Subbasin, Yushe, Fig. 2.3
T'an-Tsun, Ch'ü-Tse-Wa (Tancun, Sijiawa)	Tancun Subbasin, Yushe, Fig. 2.3
Ni-Ho-Tsun, Huang-Shih-Kou (Zhongnihe?)	Nihe Subbasin, Yushe, Fig. 2.3

R. H. Tedford, Z.-X. Qiu, L. J. Flynn (eds.), *Late Cenozoic Yushe Basin, Shanxi Province, China: Geology and Fossil Mammals: Volume I: History, Geology, and Magnetostratigraphy*, Vertebrate Paleobiology and Paleoanthropology, DOI: 10.1007/978-90-481-8714-0, © Springer Science+Business Media Dordrecht 2013

Appendix II
Licent's Trips to Southeastern Shanxi

1934 (*vide supra*, *Fig. 2.7, route A*):

June 8: Beijing→, June 9: Shijiazhuang→, June 10: Yangquan→, June 12: Pingding→, June 16: Duzhaung (S of Xiyang)→, **June 26: Lintou→, July 3: Zhangcungou→, July 22: Qiubei [Gezuitou], return to Zhangcungou→, July 27, Chen ts'ounn (Shencun →), Aug. 2: Lintou →, Aug. 8: Yuncu→**, Aug. 11: Qin Xian County→, Aug. 12: Siting→, Aug. 14: Changzhi→, Aug. 18: Licheng County→, Sept. 1: Changzhi→, Sept. 8: Jincheng→, Sept.11: Jiaozuo (Henan Province)→, Sept. 24: Shijiazhuang (Hebei Province)→, Sept. 26: Tianjin.

1935:

May 22: Tianjin→, May 23: Yuci→, **May 29: Yuncu→, May 30: Malan→, May 31: Zhaozhuang→, June 3: Gaozhuang→ June 7: Haiyan→, June 8: Zhangcungou→ , June 9: Lintou→**, June 10; Gucheng (former Wuxiang County seat)→, June 13: Siting→, June 14: Changzhi→, July 22: Huo Xian County→, July 31: Yuci.

R. H. Tedford, Z.-X. Qiu, L. J. Flynn (eds.), *Late Cenozoic Yushe Basin, Shanxi Province, China: Geology and Fossil Mammals: Volume I: History, Geology, and Magnetostratigraphy,* Vertebrate Paleobiology and Paleoanthropology, DOI: 10.1007/978-90-481-8714-0, © Springer Science+Business Media Dordrecht 2013

Appendix III
Excerpts from Licent's Diary (Vide supra, Fig. 2.7, Route B)

We are deeply indebted to Dr. K. Schmitz-Moormann for discovery of Licent's diaries in the Jesuit library in Lille, France, and for arranging for a microfilm. The microfilm and a photocopy of it are in the files of the TMNH, Tianjin. Photocopies from the microfilm are also kept in the Osborn Library of the AMNH, New York, and the IVPP, Beijing. Mr. Ray C. Gooris (originally of the AMNH Frick Laboratory) kindly translated the diary of 1934, as much as can be read. We owe him a great deal of gratitude for this and other services rendered during his retirement years. Qiu Zhan-Xiang added the diary of 1935 and augmented the translation as much as he can. Dr. Dong Wei went over and corrected the whole text.

The orthography used in Licent's diary for place names was a phonetic French transliteration of the spoken Chinese Shanxi dialect which leads to some difficulties in correlation with the names in the Wade-Giles system used in the account of Licent's Yushe trip published in 1935. To make matters more difficult, Licent was sometimes not consistent in transliteration of the same name from day to day. Utilizing the Chinese Phonetic Alphabet currently used in China, these names are given in [square brackets]. In later years, some of these places were registered as localities with serial numbers in the *Huang Ho Pai Ho* Museum, Tianjin (see Appendix IV).

1934

June 26, Tues.: 6:18 pm, entered Nan mao hoei [Nanmahui]; 7:00 pm, Church at Ling t'eou [Lintou].

June 27, Wed.: went eastward, through fossiliferous deposits, to Chang ma hoei [Beimahui]. Many turtle and *Planorbis* shells in fossiliferous sands. No large pieces, nor pockets of concentrations of fossils. Fossils are in pure yellow sand and bands of red sand, slightly clayey, stratified.

June 28, Thu.: it is raining toward morning, until 8:00 a.m., then the sun is hot. Father Landolinus Bonekamp went to west. I sent Lien Tchang [Hao Lin-Zhong] with Hoei Chang

[?] from here to Tong fang shan [Dongfangshan, Loc. 33] for exploration of fossils (40 lys [20 km] from here, and 5 lys (2.5 km) NE of Koan hoa [Guanhe, now in Guanhe Reservoir]. Lu cull ….with Hoei Tchang: elephant vertebra, rhinoceros rib, canid skull, *Gazella* jaws, turtle, horse teeth, deer and roe-deer (?) antlers, 1 rodent mandible, …. turtle shell, and two horse teeth. The fossils are always from the thick yellow sands, at about mid-section. ….

June 29, Fri.: at the 1st site (Dongfangshan) Lien Tchaung brought back two bones (1) and from the [?]2nd (Ts'ang ning [Changyin, Loc. 3])… a collection, of which 1 genus [?] of elephant (2). Lu Cull also found good things. Father L. Bonekamp returned from Yunn Tchoua [Yuncu, Loc. 51], 20 lys [10 km] west from here and 8 lys [4 km] from Kao Tchoang [Gaozhuang, Loc. 15], where also fossils were explored. ..Lu Cull brought back a number of fossils from a new site in NE (3).

June 30, Sat.: Lu Cull brought back a fragmentary turtle, but, he says, complete, with another bone (1). …I have sent Lien Tchoung with Hoei Tchang to Kao Tchoang [Gaozhuang] to see whether fossils can be excavated.

July 1, Sun.: [text unreadable].

July 2, Mon.: Lu Cull brought back numbers of antiques (1–4). ….. Lu Cull found a site full with shells, no fossils, of *Unio* and Unodonte [?] with pottery (5)….. I have sent Lien Tchoung to Cheu pei [Shibei, now Donghe, Wuxiang]. He returned. Tomorrow I will depart for Cheu pei [now Donghe] to Tchang ts'oung keou [Zhangcungou, Loc. 2]. He [Hao Lin-Zhong] brought back a very curious small tooth (6).

July 3, Tue.: ……5:22 pm arrived at Tchang ts'ounn keou [Zhangcungou]…..There is an old Christian named Ho (?) known for his expertise, who excavated fossils since 40 years. He provided me with the first fossils (4).

July 4, Wed.: at Tchang ts'oun keou [Zhangcungou] I began to excavate. The old Ho…. the foot of a cliff, where are some turtles. I decided to attack this cliff, …. for

R. H. Tedford, Z.-X. Qiu, L. J. Flynn (eds.), *Late Cenozoic Yushe Basin, Shanxi Province, China: Geology and Fossil Mammals: Volume I: History, Geology, and Magnetostratigraphy*, Vertebrate Paleobiology and Paleoanthropology, DOI: 10.1007/978-90-481-8714-0, © Springer Science+Business Media Dordrecht 2013

skull; the cliff has already been harvested for fossils in the past. Having excavated trenches over quite a long distance, ... reaches the fossiliferous level. The fossils are from the red and reddish consolidated sand, often sheathed by gray sandstone. (see plan on p. 67 [Fig. 2]). Some good bones were found (1). I have also purchased some teeth from a traveling collector (2), also from a villager......Lu Cull had a collection (3). Toward 7:00 pm, we found a big scapula in yellow sand (4). Lien tchuang [Hao Lin Zhong] went to visit a site,.... more downstream. A rib of horse (5).

July 5, Thu.: .... P. Trassaert arrived at 10:30 am......

July 6, Fri.: ...a quarry, a big block of consolidated marl,... excavate the coarse yellow sand....The fossils are....fragmentary or small (astragalus, teeth etc.). All = (1). Purchased two mastodont molars (2) found in Szen houo ning [?] 10 lys [5 km] S from here. $1.6. (3) ?.........(4)...rhinos, from the quarry.

July 7, Sat.: ....purchased from traveling collector of July 4th numbers (1). (2) Turtles and mandible of horse of Ling t'eou [Lintou, Loc. 19], brought by a man and discovered near the place of shells. No 2 of July 3rd. People continue searching. ....

July 8, Sun.: section (layers are horizontal): Loam: 0.5 m; red sand (sloping). Marl: 4.5 m; yellow sand: 3.5 m; marl...: 1 m; marl...: 1.5 m; fossiliferous yellow sand with 3 excavated galleries; red sandstone, poorly consolidated; gray sandstone. (1) Collection mixed. (2) Mandible of rhino, excavated from quarry. (3) Another mandible of rhino, as above. (4) Others found from quarry. (5) Remains mixed and other small bone from a site ..., found by Lu Cull.

July 9, Mon.: ... (1) Purchased .. of elephant, molar of similar elephant, two astragali, provenance: Tong tchoang [Dongzhuang, Loc. 32], the 2nd village upstream from Ling t'eoa [Lintou]. (2) Bone with two condyles, from quarry. (3) Collection of Lu Cull in SW. (4) Rhino from quarry, long bone. (5) Purchased bone from quarry.... (6) A small specimen from another trip W of the mountain crest..... Certainly, it seems......, in this country the fossils are disseminated, seldom whole, never grouped. We found very few skulls. Never have we found skeletons.

July 10, Tue.: we have found a new quarry, facing the 1st, in the talus of cultured terrace. Some fragmentary and friable bones. Some good pieces at the depth of 5 or 6 feet (1). From the 1st quarry: (2) vertebrae of elephant; (3) one of foot bones of elephant; (4) the same; (5) other bones; (6) rib of elephant; (7) ...elephant; (8) ...bone. See plan, p. 67. (9) bone from new quarry.....

July 11, Wed.: (1) [?]..... (2) Debris of turtle; (3) other bones of which remains of deer antler; (5) elephant of young age, *namadicus*? or mastodont? The part in front of the two molars of this mandible is worn in a scoop. The part behind is garnished with crests formed by small mammillae of 1–2 mm.... We found at Tchang ts'oua keou [Zhangcungou] 2 or 3 parts in front of ... teeth. Perhaps a young mastodont. Vide inf.; (4) great quarry No. 1.

July 12, Thu.: .... (1) From 1st quarry, in gallery...... (2) quarry No._ yesterday; At Ling t'eao [Lintou] people have excavated at five places to find what I designated yesterday. The jaw of elephant under the block of gravels fallen from...... (3) atlas of elephant from quarry 1 in Tchang ts'ounn keou [Zhangcungou]; (4) other bones, as above; (5) quarry No. 3.......

July 13, Fri.: (1) quarry 3. We have not found any skull anywhere and I have heard talk of about three or four skulls found in this place; (2) ...vertebrae of elephant from quarry 3; (3) other bones from a quarry.

July 14, Sat.: (1) quarry 3. We did not work at quarry 1... (2) Elephant bone from quarry 3; (3) elephant rib, as above. (4) radius and tibia of elephant, quarry 3; (5) elephant vertebra; (6) molar of elephant (*namadicus*?),.....

July 15, Sun.: storm....

July 16, Mon.: (1) quarry 3, various bones; (2) quarry 3, big bone; (3) ... and isolated bones brought back from Ling t'eao by Lu Cull, downstream of the village.....

July 17, Tue.: ..... (1) Quarry 3, various bones; (2) big bone, quarry 3...

July 18, Wed.: I purchased a small pagoda from the brick workers.....

July 19, Thu.: we found an elephant tusk in the upper yellow sand at the border of the quarry 1.... (1) fossils from quarry 3; (2) fossils changed against debris; (3) with 1.

July 20, Fri.: ....the tusk is broken on both sides. Its total...is 245 cm. The tusk, in total, may be 3.05 m. Old Ho excavated (and purchased at very low price) 200 elephant tusks from this country. (1) Fossils found by Lu Cull from the crest W of quarry 1. (2) Fossils found by Old Ho, ibid. (3) Teeth brought by children of Sou kia keou [Sujiagou, Wuxiang] 10 lys [5 km] W. (4) Nearby quarry. (5) Digits of elephant from Sou kia keou [Sujiagou]. (6) Horse incisors appeared [?] at two points from the debris of the covering field (quarry 1). (7) Fragments of the two ends of the tusk from quarry 1 (the tusk bearing N:), 128 livres [1 livre = 490 g]. (8) Yellow sand under the marl of the section in quarry 1.

July 21, **Sat.**: (1) under the tusk in the yellow sand, quarry 1—jaw of rhino. (2) Big shells perforated [?], 1 ly [0.5 km] downstream, in a shell concentration, left river bank. Excursion along the valley to W, with Old Ho. The sediment forms a level of grayish marl where small black bones, a little rounded, are abundant (3). This level is probably that of the upper marl in the section of p. 63...... (4) Small bones.... quarry 3...... (5) Antlers, without skull, found in the gully of the valley from Tchang ts'ounn [Zhangcun, Loc. 1] to Kou tse t'eou [Gezhuitou] .... Yellow sand. (6) Horse jaw, elephant teeth, etc., ... from Cheu pei [Donghe].

July 22, **Sun.**: travel to a deposit with fish to S.....Fish is from a soft marl,...some *Planorbis*....(1). (2) Purchased fossils from marl: 1 tusk of young elephant, Kö tze t'eou [Gezuitou, Wuxiang, ~1 km W of Qiuyuan].... (3) Elephant vertebra, on the bank [?] near fish [fossils].

July 23, **Mon.**: ... (1) Collection of Old Ho, W of quarry 3, ... yellow sand. (2) Yellow sand under the marl, quarry 1. (3) Yellow sand under the lower marl, in quarry 1. (4) Shells on the cliff, down stream. ...... (5) Fish brought back by Trassaert. (6) Work at quarry 1: bone of elephant foot; rodents, turtles, etc., lower yellow sand. (7) Elephant atlas?, as above.

July 24, **Tue.**: .... (1) Bones purchased yesterday at Cheu pei [Donghe]. (2) Lower yellow sand of quarry 1. (3) Fossils purchased from Hou ma [probably Houmu, now Gengxiu], 50 lys [25 km] NE from here. (4) Ramus of *Aceratherium* from lower yellow sand, sediment 1.

July 25, **Wed.**: I attacked the beds with small black bones (No. 3 of 21st and No. 1 of 20th) on an apron. We just arrived at a quarry of a few meters in the open: quarry 4. Marl quite sandy, very wet, with small black bones (1)... Marl very homogeneous, where Trassaert found shells, fish, mammal bones, algae, etc. (2). [artifacts].... (4) Collection by local people, along the right bank of the river from Tsang ts'ounn keou [Zhangcungou], came most probably from the beds with small black bones. (5) Fossils from the section A. (6) From quarry 1, west part, upper yellow sand, we have found series of elephant ribs, very friable. Trassaert found little things at quarry 4, no big pieces, some small bones (7). Lu cull brought back some fish from Cheu pei [Donghe] .....

July 26, **Thu.**: (1) fossil canon bone transformed into pipe, found along right bank of Tchang ts'unn keou [Zhangcungou]. (2) 3 lys [1.5 km] down stream from here, from bed 4, very low on the base of the gully, white marly sand. People say there are lots of fossils in Lou tao yu [Louzeyu], 2 lys [1 km] to SW from here. (3) Turtles from the same bed. (4) Discovery by Trassaert in the same point as yesterday (No. 2). Charcoal. (5) Discovery by Lu Cull, S of quarry 4 (bed with small black bones). (6) Collection on 3 or 4 places in a gully W of the quarries....

July 27, **Fri.**: (1) purchased elephant foot-bones, and others, found at Ni ma [Yimen?, Wuxiang], 15 lys [7.5 km] SW. Fish vertebrae of 3 cm [?] in diameter. Many fossils in Ni ma [Yimen?, Wuxiang, 5 km S of Louzeyu]. We have packed for going to Kao tchoang [Gaozhuang, Loc. 16]............ (2) Teeth of a young mastodont, found along the right bank of the Tchang ts'ounn keou [Zhangcungou] valley. (3) Skull of a bear? (4) Sheep skull. The two numbers are from Pai Hai ze [Baihai, Loc. 26], 18 lys [9 km] N of Tchang ts'ounn keou [Zhangcungou]. (5) K'iou yuan [Qiuyuan], 10 lys [5 km] to the east. Depart for Chen ts'ouna (Chang ts'ounn) [Zhangcun]. 3:07: Tchang ts'ouna [Zhangcun, Loc. 1]; 3:27: Lao ti yu [Louzeyu?]; 5:30: to the village Pa tchang keou [Bazhanggou, 2.5 km SE of Yuncu]. 6:20: Village, right bank of river, Pai hai tze [Baihai, Loc. 26]. 6:53: Tcheou tchoang [?Chongchuan, Loc. 43]. 7:15; Chen ts'ouna (true pronounciation) = Chang ts'ouna [Chongchuan]....

July 29, **Sun.**: ......(1) Tiles brought back by Lu cull..... One has prepared 200 livres of "long kou" [dragon bones]. 5 lys [2.5 km] SE, Nan tchang keou [Nanzhaunggou, Loc. 24]. The fossils are obtained from here. (2) Lu Cull from East. (3) From Pai hai tze [Baihai], two good skulls of carnovores. ...... Ning kiao [Yinjiao?], 15 lys [7.5 km] in E: big exploitation of "long kou." (4) Two enormous vertebrae of elephant, obtained in west. (5) Teeth from the same beds... (7) We purchased, through the negotiation of Lei hoei tchang [Hao Lin-Zhong], ... of Tzen t'ang [?] two molars and two tusks from the same side which I visited this morning. It is affirmed of the same individial...... The first tusk.... (8) Is...solid, the other (9) Is ...more fragmentary.... The molars (10) are still unerupted, from reddish marl.... (11) Vase given by Lei hoei tchang [Hao Lin-Zhong]. (12) Fossil shells found with the tusks and molars... (13) Vases... (14) Two elephant molars purchased by Lei hoei tchang [Hao Lin-Zhong] (one upper and one lower of the same skull). (15) Two other molars purchased by lei hoei tchang [Hao Lin-Zhong] from Hai yen [Haiyan, Loc. 6], Song ye keou [Songyangou, in the gully of Haiyan], NW of Yunn Tchuang [Yuncu]. (16) Sheep skull, from Pai hai tze [Baihai]. (17) Other fossils: two large horse [hooves] from Hai yen [Haiyan]. (18) Enormous head of femur from Ma lan [Damalan], 2 lys [1 km] NE of Yunn Tchoang [Yuncu], long bones etc. (19) Big long bones from Hai yen [Haiyan, Loc. 6]. (20) From Teou yao [Taoyang, Loc. 14], 15 lys [7.5 km] E of Yunn tchoang [Yuncu]. All the bones are purchased by Lei hoei tchang [Hao Lin-Zhong], to whom the fossils are brought by the collectors.... (21) Small flat antler, found in Yunn tchoang [Yuncu], brought by children of Lei hoei tchang [Hao Lin-Zhong].

July 30, Mon.: ...... A person from Yunn tchoang [Yuncu], the village of Lei hoei tchang [Hao Lin-Zhong], proposes a good tooth of cf. *namadicus* for $3. It was purchased by $5... (1) Fossils from the west by Lu cull, yesterday evening. (2) Collection from the big valley downstream, *Helix* from loess, bones from red sand. (3) Neolithic objects. (4) ...rhinoceros from Yunn tchoang [Yuncu]...... (6) ..... Ho linn keou [Heilingou, a gully east of Lintou] is the gully which descends from the place of the elephant jaw (No 5, July 11th)...... (7) Bones from the place of the elephant jaw. (8) Bones from the site with shells, No 1 of July 3rd...... (9) Big bone of elephant...... (10) Other bones.....

July 31, Tue.: ...... (1) Foot of an equid from ling teou [Lintou]. (2) Discovery at the quarry this morning, not rich. (3) Purchased from a traveling merchant. (4) A good canine of a carnivore found during the afternoon excavation. (5) Elephant bone purchased from a person of Kö tze teou [Gezhuitou].

Aug. 1, Wed.: we extracted numbers of jaws of ruminants (gazelle) from a piece of breccia. No 6 of last page [?Heilingou]. (1) Ma lan [Damalan], a complete deer antler. Ma lan [Damalan] is 8 lys [4 km] from here...

Aug. 2, Thu.: (1) distal end of ulna or tibia of elephant....... 4:30: Depart for Ling teou [Lintou]; 5:16: Pai hai tze [Baihai]; 6:43: Tong tchoang [Dongzhuang]; 7:26: Ling teou [Lintou].

Aug. 3, Fri.: visit to the quarry No. 6 of 31st July. The niches of the fossiliferous breccia are in the diluvial sand mixed with sterile sandstone blocks...... The harvest from the quarry is much better .. yesterday... a good skull, a good jaw, and a lot of other bones. ..... (1) Trassaert found two humeri and a stem of deer antler from the yellow sand under the loess, but they were not in situ....

Aug. 4, Sat.: ... (1) Coprolites and a suid skull found by Lei hoei tchang [Hao Lin-Zhong] yesterday on route. We have extracted a nice skull of rhinoceros (2) and a nice jaw (3) from the block of the fossiliferous breccia.

Aug. 6, Mon.: we succeeded in extracting the fossils from the day before yesterday. More skulls of rhinoceros, many gazelles. We continue to excavate upstream, but only a few things (2). Neolithic objects by Lu Cull.

Aug. 7, Tue.: ... Numbering the large pieces from Ho linn keou [Heilingou]: (1) mixture extracted from a single block of breccia. (2) Pelvis and (3) Another pelvis from the same one block. (4) Skull of rhinoceros. (5) Digits of rhinoceros, in association. (6) Large elephant bone. (7) Skull of rhinoceros. (8) Femur of rhinoceros. (9) Skull of rhinoceros. (10) Large bone. (11) Deer skull with antlers. (12) Large

humerus. (13) Long bone of elephant. (14) Mixture of a single block. (15)–(22) mixture of bones. ....

Aug. 8, Wed.: ...... (1) Large elephant bone found by Lu Cull yesterday from upstream, left bank. Trassaert brought back some teeth (1) purchased at a shop in the Yu chen hsien [Yushe County], 2 teeth of elephant (?*namadicus*), two limbs of hipparion, a flat end of elephant tooth. All are from Kinn tcheau [Qin Xian County]. ....

Aug. 9, Thu.: we left 23 boxes of fossils with Lei hoei tchang [Hao Lin-Zhong]. He will take them to the railroad.....Lei hoei tchang [Hao Lin-Zhong] purchased 30 livres of fossils (1) during our visit to Ling teou [Lintou] from Hai yen [Haiyan]....(2) Collection [?] from Pai hai tze [Baihai], downstream from Chang ts'ouna [?Shencun, Loc. 27]. ....

Aug. 10, Fri.: Depart for Lu nan fou [Lu'an, in south Shanxi].

1935

May 29, Wed.: ...... 6:00: at the left border, head of the valley where there is the sediment of Ma lan [Damalan]. 6:08: .... We saw the pagoda visited with Lei hoei tchang [Hao Lin-Zhong] the night of my arrival at Yunn tchoang [Yuncu] last year. 6:20: arrived at the church of Yunn tchoang [Yuncu]. Lei hoei tchang [Hao Lin-Zhong] isn't there.... he went to Tsai wang chan [Caiwashan, Loc. 34], 70 lys [35 km] in the west, to see a place where he was asked to see quantities of fossils.

May 30, Thu.: ... Visit to Ma lan [Damalan] afternoon....6:08: Church. Purchased: (1) a spiral horn-core of antilope, and a siphneid skull, suids (teeth), 2 vertebrae of elephant, canons, etc. Lei hoei tchang [Hao Lin-Zhong] returned this evening. He told us of Mr. K'an [Gan], who knows Father Teue (Nyström?) and sent by him (39-40 years). He lives in Tchao tchoang [Zhaozhuang], 8 lys [4 km] from Yunn tchoang [Yuncu]....He is accompanied by two men.......

May 31, Fri.: visit to Tchao tchoang [Zhaozhuang, Loc. 4]. Mr. K'an [Gan] is absent. 2 lys [1 km] west of the village, right bank of the river, excavated fossils in sand .... We passed by Pai hai tze [Baihai]....

June 1, Sat.: visit to Mr. K'an [Gan], who searches for fossils. He didn't say that he was sent, although sent by Monsieur Nyström. He doesn't know P. Teilhard.....Mr. K'an [Gan] will leave tomorrow. He purchased 30 specimens. .....Last year Mr. K'an [Gan] explored at Cheou yang [Shouyang] on Tchang-Tai [now Shijiazhuang-Taiyuan railroad]. He said he worked with Mr. Andersson elsewhere. ....

June 2, **Sun**.: visit to a valley in northwest. No fossils.... (1)–(2) Shells... (3) Siphneid from red deposits.

June 3, **Mon**.: ... We continue to purchase fossils. .... A good band of marl, at the mid-height, left bank. This is at Kao tchoang [Gaozhuang, Loc. 16]. (1) Chaotic mixture of fossils. 3:27: Nan tchang keou [Nanzhuanggou, Loc. 24]..... In the evening a person brought from 8 lys [4 km] NW, Chenn hi'an [Qin Xian?], Tzeu ho [?], Tanankeou [?], a superb jaw of elephant (mastodont group) ..... (1) Chaotic mixture of fossils from sand at the site of small panorama [photo] 366, 3, 4, 5, 6. (2) Axe...

June 4, **Tue**.:... Return to Tch'ang keou [Zhangcungou?], then take panorama pictures of yesterday....... 2:26; Arrived at Liou houng keou [Luhonggou, a gully NW of Taoyang, Loc. 14]....... 2:36: ... Road from Ling teou [Lintou] to Yunn tch. [Yuncu]. 2:47: Tsin kia keou [Qiaojiagou?]... Film 123, 1, 2, Long kou [Dragon bones] (1).3:47: Left bank of the river of Pai hai [Baihai].... (2) ... brought from Chang tchuan [Chongchuancun, Loc. 43], N of ... (Ta ping keou [Taipinggou, Loc. 45]), 20 lys [10km] from Ling teou [Lintou]. (3) Sandstone concretion. Lei hoei tchang [Hao Lin-Zhong] returned from Chen pan [?] at the foot of Ho shan [Huoshan County, 180 km SW of Yushe County] 70 lys [35 km] to the west... He brought back one elephant mandible, two tusks ....People brought back one deer antler fro Pei ts'oun [Beicun, Loc. 52]....

June 5, **Wed**.: sent [?] Lei hoei tchang [Hao Lin-Zhong] for Tai kou [Taigu] with 10 boxes of fossils...... All together 1900 livres [~900 kg].... (1) Neolithic objects. (2) *Helix* from the reddish loess. (3) Fragments from the base of the reddish loess......

June 6, **Tue**. [**should be Thu**].: ......(1) Fragments from the same situation as yesterday No. 3... Afternoon visit to Hai yen [Haiyan, Loc. 6]. Small 367, 1 [photo]. The gully with fossils....

June 7, **Wed**. [**should be Fri**.]: ...Visit by Father Landolinus Bonekamp...

June 8, **Sat**.: visit to Tchang keou [Zhangcungou?]. Some pebbles broken [?] and manufactured (?) from the excavation of fossils at the bottom of the gully (base of loess). Lei hoei tchang [Hao Lin-Zhong] returned from Tai kou [Taigu]. He brought back (1) bis a triangular stone.... (2) Two artifacts. (3) Fragments of turtle...He brought back an elephant palate with two teeth... from Kao tchang [Gaozhuang, Loc. 16].

June 9, **Sun**.: visit to Ling teou [Lintou] with Father Bonekamp.

June 10, **Mon**.: leave for Lou ngan fou [Lu'an in southern Shanxi].

Appendix IV
Licent's Localities in Southeastern Shanxi (Vide supra, Fig. 2.7)

(*Notes*: *NL* not located; Numbers for localities outside of the Yushe Basin are in [ ])

Locality	Location
1 *Changts'un* (Zhangcun)	Zhangcun Subbasin, Wuxiang
2 *Changts'unkou* (Zhangcungou)	Zhangcun Subbasin, Wuxiang
3 *Changyints'un* (Changyin)	Tancun Subbasin, Yushe
4 *Chaochuangts'un* (Zhaozhuang)	Yuncu Subbasin, Yushe
5 *Ch'iaochiakou* (Qiaojiagou)	Yuncu Subbasin, Yushe
6 *Haiyen* (Haiyan)	Yuncu Subbasin, Yushe
7 *Hopeits'un* (Haobei)	Tancun Subbasin, Yushe
8 *Hoyü* (Heyu)	Yuncu Subbasin, Yushe
9 *Hsiachuangts'un* (Xiachuang)	Zhangcun Subbasin, Wuxiang
10 *Hsiaochuang* (Xiaozhuang)	Zhangcun Subbasin, Wuxiang
11 (=10) *Hsiaochuangts'un* (Xiaozhuang)	Zhangcun Subbasin, Wuxiang
12 *Hsihots'un* (NL)	?
13 *Hsinchianao* (Xinjianao, NL)	?
14 *Hsingyangts'un* (Taoyang)	Yuncu Subbasin, Yushe
15 (=4) *Ichuangts'un* (Zhaozhuang)	Yuncu Subbasin, Yushe
16 *Kaochuang* (Gaozhuang)	Yuncu Subbasin, Yushe
17 (=16) *Kaochuangts'un* (Gaozhuang)	Yuncu Subbasin, Yushe
18 (=14) *Laohsiangts'un* (Taoyang)	Yuncu Subbasin, Yushe
19 *Lingt'ao* (Lintou)	Yuncu Subbasin, Yushe
20 *Liyüts'un* (Liyu)	Tancun Subbasin, Yushe
21 *Maichangts'un* (Maizhangcun, NL)	?
22 *Malants'un* (Damalan)	Yuncu Subbasin, Yushe
23 *Miaolingshan* (Miaoling)	Yuncu Subbasin, Yushe
24 *Nanchuangkou* (Nanzhuanggou)	Yuncu Subbasin, Yushe

(continued)

R. H. Tedford, Z.-X. Qiu, L. J. Flynn (eds.), *Late Cenozoic Yushe Basin, Shanxi Province, China: Geology and Fossil Mammals: Volume I: History, Geology, and Magnetostratigraphy,* Vertebrate Paleobiology and Paleoanthropology, DOI: 10.1007/978-90-481-8714-0, © Springer Science+Business Media Dordrecht 2013

Locality

25 *Nihots'un* (Zhongnihe)

26 *Peihaits'un* (Beihai)

27 *Shents'un* (Shencun)

28 *Shipi* (Shibi)

29 *Sunchiachuang* (Sunjiazhuang, NL)

30 *T'aichüts'un* (Taiqu)

31 *Tanankouts'un* (Danangou)

32 *Tungchuangts'un* (Dongzhuang)

33 *Tungfangshants'un* (Dongfangshan)

34 *T'saiwashan* (Caiwashan)

35 *Wangchiakou* (Wangjiagou)

36 *Yinchangts'un* (Yinzhangcun, NL)

37 *Yuant'sing* (?, NL)

38 *Yushe Hsien* (Yushe County seat)

40 *Koutits'un* (?, NL)

41 *Peilits'un* (?, NL)

42 *Shenchiakou* (?, NL)

43 *Shenchuants'un* (Chongchuancun)

44 (=14) *Taoyangts'un* (Taoyang)

45 *T'aipingkou* (Dapinggou)

46 *Kingyangpingts'un* (Qingyangping)

[47] *Wuhsiang* (Wuxiang)

48 *Yinchiao* (Yinjiao)

[49] *Th'aments'un* (Yimen)

50 *Peihsiangts'un* (?, NL)

51 *Yunchow* (Yuncu)

52 *Peits'un* (Beicun)

53 *Huchiakou* (Hujiagou, NL)

55 *Kingchangts'un* (Jinzhang)

[56] *Tochiaokouts'un* (Doujiaogou)

57 *Hsuhochangts'un* (?Nihezhang)

58 *Iliuhots'un* (Niliuhe)

[59] *Tunganchuangts'un* (?, NL)

60 (=20) *Lit'ats'un* (Liyu)

61 (=25?) *Hoanhots'un* (?Zhongnihe)

62 *Changwakou* (Zhangwagou)

[63 (=49)] *Yiments'un* (Yimen)

64 *Yenliangts'un* (Yanliang)

Location

Nihe Subbasin, Yushe

Yuncu Subbasin, Yushe

Yuncu Subbasin, Yushe

Zhangcun Subbasin, Wuxiang

?

Tancun Subbasin, Yushe

Tancun Subbasin, Yushe

Yuncu Subbasin, Yushe

Tancun Subbasin, Yushe

Nihe Subbasin, Yushe

Yuncu Subbasin, Yushe

?

?

Tancun Subbasin

?

?

?

Yuncu Subbasin, Yushe

Yuncu Subbasin, Yushe

Yuncu Subbasin, Yushe

Yuncu Subbasin, Yushe

Wuxiang County seat

Nihe Subbasin, Yushe

Wuxiang, ~5 km SW of Louzeyu

?

Yuncu Subbasin, Yushe

Yuncu Subbasin, Yushe

?Yuncu Subbasin, Yushe

Yuncu Subbasin, Yushe

Wuxiang, ?NE of county seat

Nihe Subbasin, Yushe

Nihe Subbasin, Yushe

?Wuxiang

Tancun Subbasin, Yushe

Nihe Subbasin, Yushe

Yuncu Subbasin, Yushe

Wuxiang

Yuncu Subbasin, Yushe

(continued)

Locality	Location
65 *Hsichuang* (?, NL)	?
66 *Hohsits'un* (?, NL)	?
67 *Mientsekou* (?, NL)	?
68 *Changkou* (?, NL)	?

Appendix V
Excerpts from the Correspondence of E. Nyström and C. Frick

(Drawn mostly from the Frick Archives at the American Museum of Natural History, New York)

Oct. 31, 1932, Nyström to Frick: "Since in your letter of Sept. 24th you express a wish to have the bone-expert "Buckshot" in the field as well, and since the natives at PaoTe [Baode] do not seem so keen to work this time, I have decided to send out "Buckshot" as well and trust it meets your approval. But Granger, who knows both Liu [Liu Xigu] and "Buckshot" [Gan Quanbao] very well, was of the opinion that they should not work in the same place. So I think I will send him elsewhere in Shansi. I called him round here the other day and he stated that he was willing to go and to make sure of it. I handed him some money for outfit, such as warm clothes, acetylene torches....As for his destination I know a place in SE Shansi, where my assistants discovered dragon bone pits and from where they brought some specimens. But this is long ago and the specimens were not very complete.

Jan. 11, 1933, Nyström to Frick: as for "Buckshot", he returned a few weeks ago. You will recall that as an experiment I send him to a brand-new field in SE Shansi, discovered by the assistants of my research institute in 1928. "Buckshot's" instructions were to proceed to this field, but on the way he should spend some time in the famous dragon bone market-place, Ch'i Chou, south of Pao Ting [Baoding] in this province (Hopei).... "Buckshot" heard that especially at the town Yu She Hsien [Yushe County] fossil skulls had been excavated.

Jan. 27, 1935, Nyström to Frick: ....Buckshot left within a fortnight of my telegram of Dec. 20th and I have already received a letter of Jan. 13th from SE Shansi signed by him and his clerk [Liu Deshun]. ....Before he went, P. Teilhard de Chardin called here and gave me interesting information of the district of Yu She, about 60 miles south of our present

fossil localities in SE Shansi. Teilhard said that in this district many "dragon bones" have recently been found and quite an export has taken place of teeth of various kinds, even as far as Shanghai. I gave Buckshot urgent instruction not to fail to proceed to Yu She and the localities indicated by Teilhard, and try to get as much material he could, leaving his clerk to manage in the meantime the excavation in the old places.

June 15, 1935, Nyström to Frick: .....2) The "Buckshot" party. To SE Shansi. This party has been away about 5 months, having left in the beginning of January. You will remember that P. Teilhard de Chardin proposed that Buckshot should pay a visit to Yu She Hsien in SE Shansi.... After working some time in the Shou Yang [Shouyang] localities.....with his clerk Liu Te Shun, they both started for Yu She and I must say that they had great success. In the meantime I have found on a Chinese map that this district has always been famous for "dragon bones." There is only one snag in the way. A Catholic Father Licent has been working there for his museum in Tientsin [Tianjin], the Huang Ho – Pei Ho Museum, buying up dragon bones and he seemed to resent that Buckshot intruded into "his" territory. Buckshot and myself [sic] have been discussing the situation and come to the conclusion that Licent cannot always be there and as soon as he leaves the field will be free. This, Buckshot thinks, will occur next autumn or winter. You will see from the fossil lists that even in spite of the opposition show, Buckshot and Liu Te Shun have acquired for you quite a shipment of interesting specimens.

Apr. 7, 1936, Nyström to Frick: .... Buckshot and his clerk Liu Te Shun returned recently from SE Shansi. As you may have read in the papers, the communist party entered Shansi...... the village head-men advised Buckshot and his clerk to refrain from work for the time being....

R. H. Tedford, Z.-X. Qiu, L. J. Flynn (eds.), *Late Cenozoic Yushe Basin, Shanxi Province, China: Geology and Fossil Mammals: Volume I: History, Geology, and Magnetostratigraphy,* Vertebrate Paleobiology and Paleoanthropology, DOI: 10.1007/978-90-481-8714-0, © Springer Science+Business Media Dordrecht 2013

Sept. 27, 1937, Nyström to Frick: only Buckshot is here just now. He made his way to Peiping [Beijing] through the Western Hills and left four cases of fossils up in Shansi and three down at Shih Chia Chuang [Shijiazhuang] with a friend… Buckshot's assistant Liu Te Shun is still in SE Shansi, but the work is rather at a standstill there just now….

Nov. 25, 1939, Secretary of Frick to Nyström: …It is self evident that under present circumstances it is impossible to carry on satisfactory field work and that the parties are having to mark time. Because of the general pessimistic outlook, Mr. Frick has decided to advise you to discontinue the work for the present…

Appendix VI
Quan-Bao Gan's Collection from Yushe

1935: Jan.–May: 170 specimens;
1936: Feb.–Mar: 14 specimens; Jun.–Sept.: 42 specimens.

Locality (Fig. 2.3)	Number of fossil specimens	
Gao Chuang (Gaozhuang)	3	Equid foot bone, cat mandible, elephant teeth
Chao Chuang (Chongchuan)	68	Cat, deer, goat, skulls and jaws
Chang Wa kou (Zhangwagou)	54	Cat, deer, goat, horse, elephant skulls, jaws, teeth, bones
Niu Wa kou (Niuwagou)	19	Cat, pig, goat, elephant skulls, jaws and teeth
Nanchuankou (Nanzhuanggou)	18	Cat, pig, deer, goat, rodent, elephant, bird, skulls, jaws, antlers, teeth and bones
Ma Tzu kou (Mazegou)	4	Cat jaws, horse limb-bones
Ta Nan kou (Danangou)	5	Turtles, deer skulls with antlers
Pai Hai tsun (Baihai)	11	Cat skulls
Hai Yen tsun (Haiyan)	2	Cat skull, deer antler
Xia Kou (Xiakoucun)	42	Cat, horse, pig, deer, goat, rodent, skulls and jaws
	226 (total)	

R. H. Tedford, Z.-X. Qiu, L. J. Flynn (eds.), *Late Cenozoic Yushe Basin, Shanxi Province, China: Geology and Fossil Mammals: Volume I: History, Geology, and Magnetostratigraphy*, Vertebrate Paleobiology and Paleoanthropology, DOI: 10.1007/978-90-481-8714-0, © Springer Science+Business Media Dordrecht 2013

Appendix VII
Some of Zhan-Xiang Qiu's Collection from the Yushe Area, 1979–1981

QY	Locality (see Fig. 2.3)	Fossils
34	S of Zhangcun	Hipparionine teeth
37–44	Yinjiao	Molars of *Stegodon, Anancus*, hipparionine teeth
45–49	Sangjiagou	Hipparionine teeth
51–57	Ouniwa	Teeth of various animals
66–67	Yaergou, Gaozhuang	Snout of *Paracamelus*, hyaenid jaw
69–71	Shencun	Canid skull, rodents
72–76	Zhaozhuang	Teeth of various animals
81–87	Danangou	Teeth of various animals
96	Damalan	Rodents
97	Qizigou, Zhaozhuang	Rodents
103–105	Haiyan	Hipparionine teeth
107–108	Liupinggou, Haiyan	*Equus* teeth
111–115	Shunnangou, Danangou	Teeth of various animals
133	Guanshang	Suid teeth
135	Wangning	Deer antlers
137	Qingyangping	Teeth of *Equus*, ?*Bison*

R. H. Tedford, Z.-X. Qiu, L. J. Flynn (eds.), *Late Cenozoic Yushe Basin, Shanxi Province, China: Geology and Fossil Mammals: Volume I: History, Geology, and Magnetostratigraphy*, Vertebrate Paleobiology and Paleoanthropology, DOI: 10.1007/978-90-481-8714-0, © Springer Science+Business Media Dordrecht 2013

Appendix VIII
Columnar Sections Measured by the Sino-American Yushe Project

A. Yuncu Subbasin

Abbreviation	Location	Year	Abbreviation	Location	Year
CAN	Nanmahui	1987	CAZ	Zhaozhuang	1988
CAB	Beimahui	1987	CAX	Xiongshugou	1988
CAL	Lintou	1987	CAXs	Xiongshugou, suppl.	1988
CALs	Lintou, supplementary	1987	CAY	Yexigou	1988
CAT	Taoyang	1988	CAM	Mishagou	1988
CAS	Shaoxianggou	1988	CACh	Changerwa	1988
CAGu	Guaigou	1988	CALj	Liujiagou	1988
CAHo	Honggou	1988	CAP	Liupinggou	1988
CAH	Shagou	1987	CAZh	Zhengzhaigou	1988
CAJ	Jingjiagou	1987	CAQ	Qingyangping	1991
CAG	Gaozhuang	1987			
CANa	Nanzhuanggou	1988	**B. Tancun Subbasin**		
CANas	Nanzhuanggou, suppl.	1988	CAW	Sijiawa	1991
CAC	Culiugou	1988	CAWl	Sijiawa rood section	1991
CACs	Culiugou, suppl.	1988	CAU	Jiayucun	1991
		(continued)	CAUl	Jiayucun, "Larry" section	1991

R. H. Tedford, Z.-X. Qiu, L. J. Flynn (eds.), *Late Cenozoic Yushe Basin, Shanxi Province, China: Geology and Fossil Mammals:*
Volume I: History, Geology, and Magnetostratigraphy, Vertebrate Paleobiology and Paleoanthropology,
DOI: 10.1007/978-90-481-8714-0, © Springer Science+Business Media Dordrecht 2013

Appendix IX
China–America Yushe Field Party Members and Contributors to the Series Late Cenozoic Yushe Basin, Shanxi Province, China: Geology and Fossil Mammals

Guan-Fang Chen

Laboratory of Paleomammalogy, Institute of Vertebrate Paleontology and Paleoanthropology, Chinese Academy of Sciences, Xizhimenwai Ave., 142, 100044, Beijing, PRC.

Eric Delson

Department of Anthropology, Lehman College/CUNY; and Division of Paleontology, American Museum of Natural History, Central Park West at 79 St., New York, NY 10024, USA.

Wei Dong

Laboratory of Paleomammalogy, Institute of Vertebrate Paleontology and Paleoanthropology, Chinese Academy of Sciences, Xizhimenwai Ave., 142, 100044, Beijing, PRC.

Will Downs (deceased)

Formerly Geology Department, Bilby Research Center, Northern Arizona University, Flagstaff, AZ 86001, USA.

Lawrence J. Flynn

Peabody Museum of Archaeology and Ethnology, and Department of Human Evolutionary Biology, Harvard University, Cambridge, MA 02138, USA.

Kainian Huang

Department of Geological Sciences, University of Florida, Gainesville, FL, 32611 USA.

Qiang Li

Laboratory of Paleomammalogy, Institute of Vertebrate Paleontology and Paleoanthropology, Chinese Academy of Sciences, Xizhimenwai Ave., 142, 100044, Beijing, PRC.

Yu-Qing Li

Tianjin Natural History Museum, Machangdao Ave., 296, Tianjin, PRC.

Michael Morlo

Abteilung der Meselforschung, Forschungsinstitüt Senckenberg, Frankfurt am Main, Germany

Neil D. Opdyke

Department of Geological Sciences, University of Florida, Gainesville, FL 32611, USA.

Zhan-Xiang Qiu (Qiu Z.-X.)

Laboratory of Paleomammalogy, Institute of Vertebrate Paleontology and Paleoanthropology, Chinese Academy of Sciences, Xizhimenwai Ave., 142, 100044, Beijing, PRC.

Zhu-Ding Qiu

Laboratory of Paleomammalogy, Institute of Vertebrate Paleontology and Paleoanthropology, Chinese Academy of Sciences, Xizhimenwai Ave., 142, 100044, Beijing, PRC.

Norbert Schmidt-Kittler

Johannes Gutenberg-Universität Mainz, Institüt für Geowissenschaften, Lehreinheit (LE) Paläontologie D-55099, Mainz, Germany

Ning Shi

Institute of Palynology and Quaternary Sciences, 37073, Göttingen, Germany

Richard H. Tedford (deceased)

Formerly Division of Paleontology, American Museum of Natural History, Central Park West at 79 St., New York, NY 10024, USA.

Xiaoming Wang

Department of Vertebrate Paleontology, Natural History Museum of Los Angeles County, 900 Exposition Blv., Los Angeles, CA 90007, USA, and Laboratory of Paleomammalogy, Institute of Vertebrate Paleontology and Paleoanthropology, Chinese Academy of Sciences, Xizhimenwai Ave., 142, 100044, Beijing, PRC.

Wen-Yu Wu

Laboratory of Paleomammalogy, Institute of Vertebrate Paleontology and Paleoanthropology, Chinese Academy of Sciences, Xizhimenwai Av., 142, 100044, Beijing, PRC.

Xiao-Feng Xu

Department of Geological Sciences, Southern Methodist University, Dallas, TX 75275, USA.

De-Fa Yan (deceased)

Formerly Laboratory of Paleomammalogy, Institute of Vertebrate Paleontology and Paleoanthropology, Chinese Academy of Sciences, Xizhimenwai Ave., 142, 100044, Beijing, PRC.

Jie Ye

Laboratory of Paleomammalogy, Institute of Vertebrate Paleontology and Paleoanthropology, Chinese Academy of Sciences, Xizhimenwai Ave., 142, 100044, Beijing, PRC.

Shao-Hua Zheng

Laboratory of Paleomammalogy, Institute of Vertebrate Paleontology and Paleoanthropology, Chinese Academy of Sciences, Xizhimenwai Ave., 142, 100044, Beijing, PRC.

Index

A

Academia Sinica, 23, 33, 69
Adcrocuta, 20
Alcicephalus, 14, 20
Allosiphneus, 14, 23
Alluvium, 21, 22, 41, 52, 59, 62, 70
American Museum of Natural History (AMNH), 2, 20, 25, 31, 69
Anancus, 13
Andersson, J. G., 11–13, 31
Antilospira, 18
Artiodactyla, 5
Atmospherically Tuned Neogene Time Scale (ATNTS), 80, 81
Axis, 14, 35, 39, 50, 56, 66

B

Bai, 12, 60
Baihai, 17, 26, 52, 59, 70
Baode, 1, 2, 11, 12, 15, 20, 39, 40, 63, 79, 80, 82
Baodean age, 81, 82
Beijing, 12, 13, 15, 16, 18, 20, 21, 23, 25, 41
Beryllium, ^{10}Be, 22, 62
Biostratigraphic, biostratigraphy, 1, 2, 4, 5, 24, 25, 33, 40, 57, 69, 76, 77, 79, 81, 82
Bison, 18, 20

C

Camelidae, 18
Camp, C. L., 20
Canis, 13, 20, 62
Canidae, 20
Caprolagus, 20
Cao J.-X., 21–23, 41
Carnivora, 5
Cavicornia, 18
Cenozoic Research Laboratory, 20, 33
Cervavitus, 12
Cervidae, 18
Cervocerus, 13, 14, 63
Chardina, 61
Chardinomys, 21
Chasmaporthetes, 14, 20, 60
Cheema, I. U., 24
Chilotherium, 14, 20, 22, 23
Chleuastochoerus, 13, 20, 63
Chron, 1, 24, 25, 39, 49, 66, 69, 70, 73–75, 77, 79–81
Clark, J. D., 24

Conglomerate, 14, 15, 18, 20, 28, 39–41, 47, 48, 50–52, 54, 56, 58–61, 63, 64, 66
Culiugou Member, 43, 52, 54, 55, 60, 81, 82
Cyprinid, 21

D

Damalan, 17, 56, 59, 70
Danangou, 28
Daqiang Formation, 41, 64
Demagnetization, 71
Dengyucun, 28, 48, 52, 54
Dicerorhinus, 13
Dinofelis, 20
Dip, 3, 16, 35, 39, 47, 50, 52, 55, 63, 64, 66, 72, 73
Dipoides, 18
Downs, Will, 4, 25–27
Dragon bones, 11–13, 20

E

Efremov, J. A., 21
Elaphurus, 39
Eospalax, 39
Eostyloceros, 13, 14, 20
Equidae, 18, 20, 21
Equus, 12, 16, 18, 40, 56, 82

F

Fen He, 7
Fish, 18, 21, 35, 40, 42, 50
Fluviatile, 8, 20–22, 41, 47, 48, 50, 52, 54, 56, 62, 69
Forstén, A., 24
Frick, C., 7, 20, 31, 76
Frick Collection, 20, 25
Fugu, 20

G

Gan Quan-Bao, 20
Gaozhuang (village), 17, 21, 25, 35, 47, 48, 51–53, 55
Gaozhuang Formation, 11, 22, 25, 26, 35, 41, 43, 50–57, 60, 62, 66, 69, 73, 75, 77, 79–81
Gaozhuangian age, 1, 14, 82
Gastropod, 50, 61
Gauss, 24, 69, 70, 73, 75, 77, 80
Gazella, 13, 14

R. H. Tedford, Z.-X. Qiu, L. J. Flynn (eds.), *Late Cenozoic Yushe Basin, Shanxi Province, China: Geology and Fossil Mammals: Volume I: History, Geology, and Magnetostratigraphy*, Vertebrate Paleobiology and Paleoanthropology, DOI: 10.1007/978-90-481-8714-0, © Springer Science+Business Media Dordrecht 2013

Gengxiu, 14, 15, 23, 39, 57, 58, 60, 63
Geological Bureau of Shanxi Province, 21, 41, 61, 62
Geological Survey of China, 1, 13, 63
Geomagnetic Polarity Time Scale (GPTS), 2, 76, 77, 79, 81, 82
Gilbert, 69, 70, 73, 75, 77, 80
Giraffidae, 14, 18, 20
Granger, W., 20

H

Haiyan, 4, 11, 17, 22, 56, 59, 70, 75
Haiyan Formation, 4, 11, 22, 24, 26, 35, 41, 43, 56–58, 60, 61, 69, 73, 77, 79–82
Haobei, 28, 48
Hebei, 11, 12, 16, 20, 40, 82
Hejin, 12
Hematite, 72, 73
Heissig, K., 24
Hiatus, 35, 39, 44, 55, 62, 64, 65, 69, 76, 77, 79–82
Hipparion, 11–14, 16, 18, 20, 21, 25, 39, 63
Hoang-ho Pei-ho. See Musée Hoang-ho Pei-ho de Tientsin
Honanotherium, 63
Houma, 13
Hounao, 21
Hyaena, 14
Hyaenidae, 14, 20, 22
Huang He, 7, 8, 12

I

Institut de Géo-biologie, 33
Institute of Vertebrate Paleontology and Paleoanthropology (IVPP), 2, 4, 12, 14, 21–25, 27, 29, 31–33, 41, 69, 76, 77
Isothermal remanent magnetization (IRM), 72

J

Jiayucun, 28, 48, 52

K

Kinafond, 39

L

Laboratory of Vertebrate Paleontology, 21, 31
Lacustrine, 1, 7, 8, 11, 14, 16, 18, 20, 21, 35, 39–42, 50, 52, 54, 56, 57, 60–62, 64, 66, 69
Lagrelius, A., 39
Lagrelius collection, 12, 31
Lake, 1, 35, 41, 42, 56, 62
Landolinus, B., 16
Leecyaena, 21
Leroy, P., 18, 20, 40
Li, C.-K., 25
Licent, E., 5, 7, 16–21, 23, 80
Licent collection, 24, 31–33
Lintou, 16–18, 25, 40, 59, 70, 76
Lithostratigraphy, 22, 39, 41, 42, 62, 70, 73
Lishi Loess, 23, 41, 62, 64, 67
Liu, S.-G., 20
Liujiagou, 26, 55, 56, 59, 70
Liutan, 28, 50
Liyücun, 48, 52, 56, 59, 70
Loess, 1, 2, 8, 14, 15, 19, 23, 35, 39, 41, 42, 45, 47, 48, 52, 55–57, 59–67, 69, 70, 74, 79, 82

Loess plateau, 8, 41, 64, 79, 82
Louzeyu Formation, 21, 23, 24, 41, 42, 62, 63, 66
Lynx, 62

M

Magnetic Polarity Time Scale (MPTS), 2, 74, 75, 77, 79
Magnetite, 72
Magnetochronology, 62
Magnetostratigraphy, 2, 4, 5, 25, 26, 28, 41, 45, 69, 70, 75, 76, 79, 80, 82
Mahui Formation, 8, 11, 22, 28, 35, 41, 43, 47, 48, 50–52, 54, 57–59, 61, 63, 64, 66, 69, 73, 75, 77, 79–81
Malan, 26, 35, 41, 57, 60–62, 64, 65, 67
Malan Loess, 35, 41, 57, 60–62, 64, 65, 67
Marl, 18, 19, 21, 22, 35, 40–42, 47, 48, 50–53, 55, 59–61, 73
Mastodon, 14
Matuyama, 24, 56, 64, 69, 70, 75, 77, 81
Mazegou Formation, 11, 22, 41, 43, 51, 52, 55, 56, 58, 60, 62, 66, 69, 73, 77, 81, 82
Mazegouan age, 1, 80, 81
Metailurus, 13
Microstonyx, 14
Microtragus, 14
Mollusc, 52, 55, 56
Moschus, 14
Musée Hoang-ho Pei-ho de Tientsin, 16, 17, 31, 33
Myospalacine, 14, 61
Myospalax, 60

N

Nanzhuanggou village, 26, 43, 51–54, 58
Nanzhuanggou Member, 43, 51–53, 58, 81
Natural remanent magnetization (NRM), 71, 77
Naturhistorisches Museum, Basel, ix
Nei Mongol, 16
Neocricetodon, 28
Neogene, 1, 2, 4, 5, 9, 22, 25, 31, 33, 35, 39, 56, 66, 69, 79–82
Nihe, 1, 3, 8, 9, 11, 39, 47, 55, 57, 60–62, 66
Nihewan, 1, 16, 40, 79, 82
Nihewanian age, 80
North China, 1, 2, 4, 7, 11, 12, 16, 22, 23, 63, 66, 77, 79, 81, 82
North China Plain, 7, 63, 66
Nyström, E. T., 20

O

Olduvai, 4, 66, 81
Ouniwa, 1, 3, 8, 9, 11, 15, 23, 28, 29, 39, 47, 48, 54, 55, 57–62, 66, 67
Ostracod, 23, 50, 52, 61

P

Paleomagnetic, paleomagnetism, 2, 7, 22, 24, 25, 35, 43, 45, 48, 51, 57, 59, 64, 69, 70, 77, 79
Pachycrocuta, 62
Palaeotragus, 14
Paramachairodus, 18, 28
Palynology, 62, 63
Pelecypod, 61
Pentalophodon, 13
Percrocuta, 20
Perissodactyla, 5
Pinyin, 30–32

Pleistocene, 1, 2, 4, 7, 16, 22, 23, 26, 28, 35, 39, 41, 56, 60, 62–64, 67, 74, 75, 77, 79–82
Plesiohipparion, 32
Pliosiphneus, 61
Pollen, 42, 63
Primates, 5
Proboscidea, 5, 15, 18
Procapreolus, 13
Propotamochoerus, 13

Q
Qin County, 13, 15, 17, 34, 39, 41
Qin Xian, 8, 13, 15–17
Qingyangping, 28, 56
Qinshui Synclinorium, 7, 9, 66

R
Raza, S. M., 24
Rhinoceros, 13, 14, 21
Rhinocerotidae, 20
Red loam, 14, 18, 39, 41, 63, 64
Renjianao Formation, 21, 41, 61, 63

S
SAYP. *See* Sino-American Yushe Project
Sericolagus, 20
Shanxi Graben, 66
Shencun, 16, 17, 55, 59, 70
Shi, N., 22–24
Shouyang, 1, 2, 13, 20
Sino-American Yushe Project (SAYP), 2, 5, 11, 24, 25, 31, 33, 77, 79
Siphneus, 14, 20, 23, 61
Siting, 8, 17, 41
Songyangou, 26
Stegodon, 14, 20, 26
Suidae, 14, 20
Synclinorium, 7–9, 16, 35, 39, 47, 66
Systematics, 1, 5, 21, 25, 26, 31–33, 77, 81, 82

T
Taigu, 2, 14, 15, 18, 21
Taihang, 7, 8, 16, 35, 39, 48, 63, 66, 67, 82
Taiqu, 28
Taiyuan, 1, 2, 8, 11, 20, 21
Tancun, 1–3, 8, 9, 11, 12, 14, 20, 25, 26, 28, 39, 41, 47–50, 52–63, 66, 70, 75, 77, 81
Taoyang (village), 32, 47, 51, 59, 66, 70, 75, 76
Taoyang Member, 43, 51–54, 57, 58, 77, 80, 81
Teilhard de Chardin, P., 5, 13–16, 18, 20, 21, 23, 33
Tianjiangou, 61
Tianjin, 15–18, 22, 31–33, 41
Tianjin Natural History Museum (TNHM), 16, 22, 24, 25, 28, 32

Tobien, H., 13, 23, 24
Tragocerus, 14
Trassaert, M., 17–20
Triassic, 35, 39–42, 47, 48, 50–52, 54–61, 63–66, 69–71, 73, 75
Tunliu, 8, 16, 64
Turtle, 15, 18, 40

U
Unconformity, 22, 41–43, 45, 48, 49, 51, 54, 56, 58, 60, 62, 73, 75
Uppsala, 12, 69
Ursidae, 20

V
Virtual Geomagnetic Pole (VGP), 43, 49, 72–75
Viverridae, 20

W
Wade-Giles, 30, 31
Wangning Formation, 22, 42, 62, 77
Wu, Wen-yu, 4, 22, 25, 27, 28, 49
Wucheng Loess, 56, 62, 64–66
Wuxiang, 1–3, 7–9, 11–18, 20, 21, 25, 31, 35, 39–41, 45, 48, 62–64, 69, 81

X
Xiangyuan, 8

Y
Yan, D.-F., 4, 22, 24, 25, 27
Young, C.-C., 13–16, 20, 33, 39
Yangia, 20
Yuci, 17, 18, 21
Yuncu, 1–4, 8, 9, 11, 16–22, 25, 26, 28, 39–44, 47, 48, 50–52, 54–62, 64, 66, 69, 70, 75–77, 79–81
Yushe County Museum, 4, 22, 23, 26–29, 31
Yushean, 1

Z
Zdansky, O., 12–14, 39
Zhangcun (village), 1, 3, 8, 11, 15–19, 21–24, 28, 31, 39, 40–42, 44, 48, 59, 61–64, 66, 69, 70, 75, 77, 79
Zhangcun Formation, 11, 18, 21–23, 41, 42, 62, 63
Zhangcungou, 15–19, 21, 28, 31, 39, 40, 42, 48, 64
Zhanghe Fault, 66
Zhangjiagou, 39, 64
Zhaozhuang (village), 32, 51
Zhongcun Basin, 64
Zhuozhang, 7, 8, 11, 35, 39, 47, 48, 50, 52, 54, 60, 63, 64, 66, 68
Zokor, 14, 23, 39, 60